Security Games

Security Games: Surveillance and Control at Mega-Events addresses the impact of "mega-events" – and especially the Olympic Games and the World Cup – on wider practices of security and surveillance. "Mega-events" pose peculiar and extensive security challenges. The overwhelming imperative is that nothing should go wrong. There are, however, an almost infinite number of things that can "go wrong"; producing the perceived need for pre-emptive risk assessments, and an expanding range of security measures, including extensive forms and levels of surveillance. These measures are delivered by a "security/industrial complex" consisting of powerful transnational corporate, governmental and military actors, eager to showcase the latest technologies and prove that they can deliver "spectacular levels of security" to protect the brand image of the country or city in question. As the authors demonstrate, these events have become occasions for experiments in monitoring people and places, and as such, important moments in the development and dispersal of surveillance, and the more routine monitoring of people and place. As the exceptional conditions of the "mega-event" become the norm, *Security Games: Surveillance and Control at Mega-Events* provides a glimpse of a possible future that is more intensively and extensively monitored.

Colin J. Bennett is professor of political science at the University of Victoria. In addition to numerous articles, he has published five books, the latest being *The Privacy Advocates: Resisting the Spread of Surveillance*. He is currently co-investigator on a large international project entitled "The New Transparency: Surveillance and Social Sorting."

Kevin D. Haggerty is editor of the *Canadian Journal of Sociology* and book review editor of the international journal *Surveillance & Society*. He is professor of sociology and criminology at the University of Alberta. Among his most recent publications is the co-edited book (with Minas Samatas) *Surveillance and Democracy*.

Security Games

Surveillance and Control at Mega-Events

Edited by Colin J. Bennett and
Kevin D. Haggerty

a GlassHouse book

First published 2011
by Routledge
2 Park Square, Milton Park, Abingdon, Oxon OX14 4RN

Simultaneously published in the USA and Canada
by Routledge
711 Third Avenue, New York, NY 10017

A GlassHouse book

Routledge is an imprint of the Taylor & Francis Group, an informa business

© 2011 editorial matter and selection: Colin J. Bennett and Kevin D. Haggerty. Individual chapters: the contributors.

The right of the editors to be identified as the authors of the editorial material, and of the authors for their individual chapters, has been asserted in accordance with sections 77 and 78 of the Copyright, Designs and Patents Act 1988.

All rights reserved. No part of this book may be reprinted or reproduced or utilised in any form or by any electronic, mechanical, or other means, now known or hereafter invented, including photocopying and recording, or in any information storage or retrieval system, without permission in writing from the publishers.

Trademark notice: Product or corporate names may be trademarks or registered trademarks, and are used only for identification and explanation without intent to infringe.

British Library Cataloguing in Publication Data
A catalogue record for this book is available from the British Library

Library of Congress Cataloging in Publication Data
Security games : surveillance and control at mega-events / edited by
   Colin J. Bennett and Kevin D. Haggerty.
      p. cm.—(A GlassHouse book)
   ISBN 978-0-415-60262-4 (hbk)—ISBN 978-0-203-82747-5 (ebk)
   1. Crowd control—Case studies. 2. Spectator control—Case studies. 3. Security systems—Case studies. I. Bennett, Colin J. (Colin John) 1955– II. Haggerty, Kevin D.
   HV8055.S43 2011
   363.32'3—dc22
                                               2010040652

ISBN: 978-0-415-60262-4 (hbk)
ISBN: 978-0-203-82747-5 (ebk)

Typeset in Times New Roman
by RefineCatch Limited, Bungay, Suffolk

Printed and bound in Great Britain by
CPI Antony Rowe, Chippenham, Wiltshire,

Library
University of Texas
at San Antonio

Contents

Preface		vii
List of contributors		ix
	Introduction: Security games: surveillance and control at mega-events COLIN J. BENNETT AND KEVIN D. HAGGERTY	1
1	**Rethinking security at the Olympics** DANIEL BERNHARD AND AARON K. MARTIN	20
2	**Olympic rings of steel: constructing security for 2012 and beyond** PETE FUSSEY AND JON COAFFEE	36
3	**Surveilling the 2004 Athens Olympics in the aftermath of 9/11: international pressures and domestic implications** MINAS SAMATAS	55
4	**The spectacle of fear: anxious mega-events and contradictions of contemporary Japanese governmentality** DAVID MURAKAMI WOOD AND KIYOSHI ABE	72
5	**"Secure Our Profits!": the FIFA™ in Germany 2006** VOLKER EICK	87
6	**Event-driven security policies and spatial control: the 2006 FIFA World Cup** STEFANIE BAASCH	103

7 **Commonalities and specificities in mega-event securitization: the example of Euro 2008 in Austria and Switzerland** FRANCISCO R. KLAUSER	**120**
8 **Gran Torino: social and security implications of the XX Winter Olympic Games** CHIARA FONIO AND GIOVANNI PISAPIA	**137**
9 **Mega-events and mega-profits: unravelling the Vancouver 2010 security–development nexus** ADAM MOLNAR AND LAUREEN SNIDER	**150**
10 **Knowledge networks: mega-events and security expertise** PHILIP BOYLE	**169**
Index	185

Preface

"Mega-events" such as the Olympic Games and the FIFA World Cup are now monumental cultural spectacles that pose peculiar and extensive security challenges. The world's attention focuses on a host nation or city, eager to project the best of impressions and thus attract future investment and tourism. The overwhelming imperative of "destination marketing" produces a pervasive sentiment that "nothing should go wrong." There are, of course, an almost infinite number of things that can "go wrong," producing extensive and pre-emptive risk assessments, and an increasing number of security measures, including wide-ranging forms and levels of surveillance.

Furthermore, organizers such as the International Olympic Committee (IOC) or the International Federation of Association Football (FIFA) now place extraordinary conditions on host cities and nations, which must make elaborate and expensive promises with respect to security. The Athens Games in 2004 had the highest documented security costs at $1.5 billion USD. Current estimates for Vancouver/Whistler suggest that around $1 billion CAD was spent on security arrangements. A "security/industrial complex" consisting of powerful transnational corporate, governmental and military actors, is eager to showcase the latest technologies, secure lucrative contracts, and deliver these various measures. At the same time, citizens tend to be more tolerant of abnormal security precautions, which they might oppose in other contexts. Local public officials often capitalize on these opportunities, treating them as an occasion to conduct real-world tests of new information systems and to implement their wish lists of measures unattainable in more normal conditions.

Mega-events have thus become opportunities for experiments in monitoring people and places, and important moments in the development and dispersal of surveillance. They are crucibles for governmental anxieties and microcosms of larger trends and processes. Through the mega-event, we can observe the complex ways that unique combinations of technology, institutional motivations, and public–private security arrangements produce security practices. In this process, the surveillance infrastructure established for one mega-event expands and becomes the standard for the future. The "Security Games" consequently have both international and domestic legacies.

There are powerful transnational forces leading to higher levels of securitization and surveillance from one event to the next, and an increasing standardization of security practices that transcend temporal, national and cultural borders. At the same time, these powerful forces are mediated by local conditions. The FIFA World Cup is not the Olympics. London is not Tokyo is not Atlanta is not Beijing is not Vancouver. Local political, administrative, cultural and economic factors also shape the experiences on the ground and the more general perceptions and legacies.

The chapters in this collection are all in one way or another concerned with this interplay between the global and the local. They focus on sporting events, rather than other international spectacles such as Expos or G8 summits. The collection offers historical and spatial comparisons through case studies of the 2004 Olympic Games in Athens, the 2006 World Cup in Germany, the 2006 Winter Olympics in Torino, the 2008 European Championships in Switzerland, the Winter Olympics in Vancouver/Whistler 2010, the future 2012 Olympic Games in London and the sequence of mega-events which have occurred in Japan since the 1964 Tokyo Olympics. The collection originates in a research workshop held in Vancouver November 20–21, 2009 just two months before the Winter Olympic Games. Scholars, policy makers and activists from many countries in Europe and North America convened to discuss questions related to the impact of mega-events on security and surveillance. This workshop was organized under the auspices of the New Transparency Initiative, a large multi-disciplinary and international project funded through the Social Sciences and Humanities Research Council of Canada.

Besides this volume, another product of this workshop was a signed declaration drawing attention to the fact that "recent Games have increasingly taken place in and contributed to a climate of fear, heightened security and surveillance; and that this has often been to the detriment of democracy, transparency and human rights, with serious implications for international, national and local norms and laws." Among other things, the declaration called for "a full, independent public assessment of the security and surveillance measures, once the Games are over, addressing their costs (financial and otherwise), their effectiveness, and lessons to be learned for future mega-events" (Vancouver Statement, 2009).

We are grateful to the authors for the professional and timely way in which they prepared drafts and responded to our editorial suggestions. We would also like to thank the other participants in the research workshop whose work does not appear in this volume. Colin Bennett would also like to thank his Australian colleagues for the stimulating intellectual environment at the Cyberspace Law and Policy Centre University of New South Wales in Sydney, where he spent time in 2010 editing this collection. We thank Bryan Sluggett, Ariane Ellerbrok and Temitope Oriola for their editorial assistance, as well as Colin Perrin from Routledge. We gratefully acknowledge financial support provided by the SSHRC Major Collaborative Research Initiative grant for "The New Transparency" project.

List of contributors

Kiyoshi Abe is professor at the School of Sociology, Kwansei Gakuin University. He is interested in the critical sociology of media and communication. In his Ph.D. dissertation he assessed and compared the academic legacy of the Frankfurt School and that of British Cultural Studies for critical media and communication studies as a means to develop new ways to interrogate the socio-political conditions of communications in contemporary information societies. Drawing on the Habermasian normative concept of the public sphere, Abe analyzed and criticized the changing political conditions of nationalism in postwar Japanese society. As a scholar who has research interests in the socio-cultural changes caused by the introduction of new information and communication technologies, he also has paid close attention to the rise and prevalence of surveillance in contemporary Japanese society. His recent research critically investigates the socio-cultural conditions of globalized surveillance society in relation to the transformation of the public sphere. His publications include "Everyday Policing in Japan: Surveillance, Media, Government and Public Opinion" in *International Sociology* 19 (2) 2004, and "The Myth of Media Interactivity: Technology, Communications and Surveillance in Japan" in *Theory, Culture & Society* 26 (2–3) 2009.

Stefanie Baasch graduated in public administration economy (Diploma 1996) and in social behavior science and history (Master of Arts in 2004). From 1996 to 2001 she worked in Civil Service organizations, which provided her with a profound knowledge of decision processes in public authorities. In the last few years she has been working as an environmental and social psychologist in interdisciplinary research projects in the field of sustainability, climate mitigation and city development. She is an expert in qualitative research methods, especially interviews, questioning, and focus groups. In 2009 she completed her Ph.D. thesis on security impacts of the FIFA Football World Cup 2006 in Germany. Currently, she works as a researcher in the field of environmental governance in climate change adaptation processes at the University of Kassel.

Colin Bennett received his Bachelor's and Master's degrees from the University of Wales, and his Ph.D. from the University of Illinois at Urbana-Champaign. Since 1986 he has taught in the Department of Political Science at the University of Victoria, where he is now Professor. From 1999 to 2000, he was a fellow at Harvard's Kennedy School of Government. In 2007 he was a Visiting Fellow at the Center for the Study of Law and Society at University of California, Berkeley. In 2010, he was Visiting Scholar at the Centre for Cyberspace Law and Policy at the University of New South Wales. His research has focused on the comparative analysis of surveillance technologies and privacy protection policies at the domestic and international levels. In addition to numerous scholarly and newspaper articles, he has published five books: *Regulating Privacy: Data Protection and Public Policy in Europe and the United States* (Cornell University Press, 1992); *Visions of Privacy: Policy Choices for the Digital Age* (University of Toronto Press, 1999, co-edited with Rebecca Grant); *The Governance of Privacy: Policy Instruments in the Digital Age* (The MIT Press, 2006 with Charles Raab); *The Privacy Advocates: Resisting the Spread of Surveillance* (The MIT Press, 2008); and *Playing the Identity Card: Surveillance, Security and Identification in Global Perspective* (Routledge, 2008 co-edited with David Lyon).

Daniel Bernhard holds a B.Sc. in International Relations and History from the London School of Economics and Political Science, and an M.Phil. in Political Thought and Intellectual History from the University of Cambridge. He is currently exploring various themes in his research, including: the concept of security, the interplay between the languages of desire and science in politics, the normativity of positive economics, and the emergence of the American libertarian movement since the 1930s. Some recent and forthcoming publications study the politics of historical memory in the Islamic world, the effects of historical knowledge in economic governance, and the politics of resistance to surveillance.

Philip Boyle is a Ph.D. candidate in the Department of Sociology at the University of Alberta (Edmonton) and currently a visiting scholar in the School of Communications at Temple University (Philadelphia). He received a B.A. in Criminal Justice from Mount Royal University (Calgary) and an M.A. in Sociology from the University of Windsor. His research interests are at the intersection of critical criminology, surveillance studies, and urban governance. His current research examines security governance at the Olympic Games and other major sporting and political events.

Jon Coaffee holds a Chair in Spatial Planning and Urban Resilience in the Centre for Urban Policy Studies, Birmingham Business School, University of Birmingham, UK. Before this appointment he worked at the Universities of Newcastle and Manchester. His works has focused upon the interplay of physical and socio-political aspects of resilience. In particular he has analyzed the ability of businesses, governments and communities to anticipate shocks, and ultimately embed resilience within everyday activities and the ability to attain a culture of "resilience." This work has been funded by a variety of UK Research Council programs for "New Security Challenges" and "Global Uncertainties." He has published widely on the social and economic future of cities, and especially the impact of terrorism and other security concerns on the functioning of urban areas. This work has been published in multiple disciplinary areas such as geography, town planning, political science, sports studies and civil engineering. Most notably he published *Terrorism Risk and the City* (Ashgate, 2003), *The Everyday Resilience of the City: How Cites Respond to Terrorism and Disaster* (Palgrave 2008) and *Terrorism Risk and the Global City: Towards Urban Resilience* (Ashgate 2009).

Volker Eick is a political scientist in the Department of Social Sciences at the Goethe Universität Frankfurt am Main, Germany and a PhD candidate in the Department of Politics at the John F. Kennedy Institute, Freie Universität Berlin, Germany. His research endeavors include the commercialization of security, new social movements, urban security regimes, and workfare. He has published in a range of academic journals such as the German journal *Bürgerrechte & Polizei/CILIP,* and in Anglophone journals such as *CITY, Contemporary Justice Review, Policing & Society, Social Justice* and *Urban Studies*. He is currently co-editing a Special Issue of *Social Justice* on "Urban Security Work Spaces" and a book on *Policing in Crisis – Policing of Crisis* (under contract with Red Quill Books, Ottawa/ON, Canada).

Chiara Fonio holds a Ph.D. in Sociology and the Methodology of Social Research. Her main research areas are surveillance studies, security and terrorism, and ICTs. As part of her doctoral thesis she carried out the first Italian qualitative research focused on surveillance cameras in the city of Milan. The book *Videosorveglianza, uno sguardo senza volto* (Video surveillance. A faceless gaze) originated from her thesis and was released in February 2007. After her empirical research she was put in charge of the surveillance camera operators' training pertaining to the social aspects and privacy implications of these technologies. Currently, she is senior researcher of ITSTIME (Italian Team for Security, Terroristic Issues and Managing Emergencies (http://www.itstime.it/). ITSTIME is based at the Università Cattolica del Sacro Cuore (Milan, Italy), Department of Sociology.

Pete Fussey is a Senior Lecturer in Criminology at the University of Essex, UK and lectures on a range of criminological themes – including criminological theory, transnational organized crime, urbanization and crime and psychological criminology – and on critical terrorism and counter-terrorism related issues. Dr Fussey's main research interest focuses on technological surveillance, particularly in relation to their deployment in relation to crime and terrorism and he has published widely on the subject. Currently, he is researching the form and impact of the 2012 Olympic security strategy and also conducting ethnographic research into organized crime and the informal economy in East London. He has also researched terrorism and counter-terrorism for a number of years and is currently working on two large-scale funded research projects looking at counter-terrorism in the UK's crowded spaces and at the future of urban resilience until 2050. Dr Fussey also has three books due for publication in 2010: *Securing the Olympic City: Reconfiguring London for 2012 and Beyond*, Ashgate (with J. Coaffee, G. Armstrong and D. Hobbs); *Terrorism and the Olympics: Lessons for 2012 and beyond*, Routledge (ed. with A. Richards and A. Silke); *Researching Crime: Approaches, Method and Application*, Palgrave (with C. Crowther-Dowey).

Kevin D. Haggerty is editor of the *Canadian Journal of Sociology* and book review editor of the international journal *Surveillance & Society*. He is Professor of Sociology and Criminology at the University of Alberta and a member of the executive team for the *New Transparency* Major Collaborative Research Initiative. He has authored, co-authored or co-edited *Policing the Risk Society* (Oxford University Press), *Making Crime Count* (University of Toronto Press), *The New Politics of Surveillance and Visibility* (University of Toronto Press) and *Surveillance and Democracy* (Routledge). Among other projects he is currently working on the *Routledge International Handbook of Surveillance Studies* (with David Lyon and Kirstie Ball).

Francisco R. Klauser is assistant professor in Political Geography at the University of Neuchâtel, Switzerland. His work focuses on the relationships between space, surveillance/risk and power, with a particular focus on public urban space and places of mobility. His research interests also include urban studies and socio-spatial theory. In recent years Francisco Klauser has developed an international portfolio of work on security and surveillance at sport mega-events, including publications on the 2006 FIFA World Cup in Germany, the 2008 Beijing Games and the 2008 European Football Championships in Switzerland/Austria. More recent empirical research also focused on the 2010 Vancouver Winter Games. He has published in a range of journals, including theme-focused contributions in several fields and languages as well as in some of the most influential Anglophone academic journals such as the *British Journal of Sociology*, *European Urban and Regional Studies* and *Environment and Planning D: Society and Space*. Francisco Klauser has also authored and co-authored two books on surveillance-related topics in German and French. Currently, he is co-editing a book on *Risk Research: Practices, Politics, Ethics* (under contract with Wiley-Blackwell), and journal special issues of *Urban Studies* on "Security, Cities and Sport Mega-Events" and of *Environment and Planning D: Society and Space* on "Revisiting Human Territoriality."

Aaron K. Martin has recently completed his Ph.D. at the London School of Economics and Political Science (LSE) in the Information Systems and Innovation Group, Department of Management. His research focuses on the political, ethical, social, and policy aspects of biometric technologies, identity systems, surveillance, and information and communication technologies. Aaron's recent academic publications include articles in the journals *Surveillance & Society* and *Communications of the Association for Information Systems*. He also works frequently with the International Development Research Centre to research privacy and surveillance aspects of information and communication technology projects in developing countries. For more information about Aaron's research, visit his website: http://personal.lse.ac.uk/martinak/.

Adam Molnar is a Ph.D. Candidate in the Department of Political Science at the University of Victoria and a research assistant for the New Transparency project. His dissertation focuses on the social, material, legal/economic and technical dimensions of interoperable security networks, particularly in the context of mega-events. He has also written on surveillance technologies and regulation in stock market reform with Laureen Snider, as well as on intersections between new materialisms, surveillance and critical security studies. Other research interests include international political sociology, cultural political economy/critical realism and critical security studies, critical security methods, and Science and Technology Studies.

Giovanni Pisapia graduated in 2002 from the Università Cattolica del Sacro Cuore (Milan, Italy) in Political Science with a thesis on "Crime in the Republic of South Africa: Reasons, Current Situation and Possible Future Solutions." In 2008 he obtained a Ph.D. in Criminology from the same university with a thesis on "The Development of a Terrorism Risk Management Framework (TRMF)." As a Consultant for UNICRI (United Nations Interregional Crime and Justice Research Institute) he designed and implemented training activities on crime prevention and justice administration. As a Security Manager at TOROC (Olympic Organizing Committee XX Olympic Winter Games – Torino 2006) he was in charge of the co-ordination and co-operation between the organizing committee and Italian law enforcement agencies for the development of the venues' security plans, the security policies, procedures and contingency plans, as well as the security transfer of knowledge document for the International Olympic Committee (IOC). As the Project Manager for the City of Johannesburg Safety and Security 2010 FIFA World Cup™, he was responsible for developing the Johannesburg Metropolitan Police Department (JMPD) 2010 FIFA World Cup™ safety and security strategy and operational plan. Currently he is responsible for developing crime mapping and crime analysis of specific areas in the City of Johannesburg.

Minas Samatas is Associate Professor of Political Sociology at the Sociology Department at the University of Crete, Greece. He has an M.A. and a Ph.D. in Sociology from the Graduate Faculty of New School for Social Research, N.Y., U.S.A. He has published articles in international journals on, among other things, "Greek McCarthyism," "Surveillance at the Athens 2004 Olympics," and "The Southern European Fortress." He is author of *Surveillance in Greece: From anticommunist to the consumer surveillance* (Pella, N.Y., 2004), and co-editor with Kevin Haggerty of *Surveillance and Democracy* (Routledge, 2010). He is also the Greek representative in the Managing Committee of the European Research Network "Living in Surveillance Societies" (LiSS), and co-ordinator of the research group "Post-authoritarian South-European surveillance societies."

Laureen Snider is a Professor of Sociology at Queen's University specializing in corporate crime, surveillance, and legal reform. Recent publications include "Framing E-Waste Regulation: The Obfuscating Role of Power," *Criminology & Public Policy* 9 (3), (2010); "Tracking Environmental Crime Through CEPA: Canada's Environment Cops or Industry's Best Friend?," with Suzanne Day and April Girard *Canadian Journal of Sociology*, 35(2) (2010); "Regulating Competition in Canada," with Suzanne Day and Jordan Watters, *Canadian Journal of Law & Society* (2009); and "Accommodating Power: The 'Common Sense' of Regulators," *Social & Legal Studies* 18 (3) (2009). She is presently working on articles for a volume titled *European Developments in Corporate Criminal Liability* (Sage, forthcoming); one on Financial Regulation for the *Annual Review of Law and Social Science*, and one on surveillance in stock market reform (with Adam Molnar).

David Murakami Wood is Canada Research Chair (Tier II) in Surveillance Studies and Associate Professor in the Department of Sociology, Queen's University, Kingston, Ontario, and a member of The Surveillance Studies Centre. He specialises in the study of the history, technologies, practices and ethics of surveillance mainly in urban contexts in international cross-cultural comparative perspective. He is also interested in: ubiquitous computing; urban resilience to disaster, war and terrorism; and science fiction. In 2009 he was Visiting Professor at the Pontifical Catholic University or Parana in Curitiba, Brazil and in 2006 he was an Exchange Visiting Fellow at Waseda University, Tokyo. He is a co-founder and Managing Editor of the international journal of surveillance studies, *Surveillance & Society*, and a co-founder and trustee of the Surveillance Studies Network (SSN). He edited the influential *Report on the Surveillance Society* for the UK Information Commissioner's Office (ICO) (2006), and has been published in a wide range of academic journals including the *European Journal of Criminology, International Relations, Society & Space, Urban Studies* and *Theory Culture and Society*. He is currently writing two books *The Watched World: Globalization and Surveillance* (Rowman & Littlefield, USA) and *Global Surveillance Societies: New Spaces of Surveillance* (Palgrave, UK).

Introduction
Security games: surveillance and control at mega-events

Colin J. Bennett and Kevin D. Haggerty

At the time of editing this collection, the 2010 FIFA World Cup was taking place in South Africa. The following was proudly proclaimed on the government's World Cup website:

> Some R665 million will be spent on procuring special equipment, including crowd-control equipment, crime scene trainers, unmanned aircraft, helicopters, 10 water cannons, 100 BMWs for highway patrol and up-to-date body armour. About 300 mobile cameras will also be used. There will be four mobile command centres at a cost of around R6 million each. These centres will feature high-tech monitoring equipment, which will be able to receive live footage from the airplanes and other cameras. These investments will continue to assist the police in their crime-fighting initiatives long after the World Cup is over.
>
> This same webpage reported that 41,000 officers from the South African Police Service would be deployed during the event, with dedicated World Cup 2010 police stations in close proximity to each of the stadiums.
>
> (South Africa, FIFA 2010)

It might be tempting to respond that South Africa, with its authoritarian history, would be expected to produce such an excessive security response. As the papers in this volume illustrate, however, there is nothing particularly remarkable about the scale of this security effort, something that is now common in host countries with more established democratic traditions. Elsewhere we have witnessed: increased use of video-surveillance at prominent and vulnerable sites; uses of secure perimeter fencing; criminal background checks for employees, volunteers and athletes; vehicle monitoring; the use of radio-frequency identification devices (RFIDs) on passes and tickets; biometric identification measures; satellite monitoring; the designation of special "fan zones" for collective viewing of events; the regulation of protest and dissent; overhead unmanned aerial vehicles; mobile fingerprinting identification systems; and enhanced controls at land, sea and air borders. Each mega-event now exhibits a "total security" effort akin to planning and deployment in times of war. Sometimes the parallel is made explicit; Chinese officials declared in February 2008 that they had entered the "combat phase" of their preparations for the Beijing Olympics (Beijing 2008).

From the vantage point of the local organizers, officials and public, "their event" is often viewed as unique. To some extent each is. But each World Cup, Olympic Games, Expo, G8 or G20 summit or other "mega-event" is also part of a broader process through which various institutions and officials learn and pass on lessons about security to their successors. International organizations such as the International Olympic Committee (IOC) or the Fédération Internationale de Football Association (FIFA) act as centralized brokers of that knowledge. Each event, therefore, tries to improve on its predecessors, and each must now deliver "spectacular security" (Boyle and Haggerty 2009), increasingly coordinated and militarized, with strong continuities across time and place. How has this situation come about, and what are the implications for our communities, our civil liberties and our democracies?

This extensive security effort might be easily justified if there was a demonstrable threat of serious incident at those events, such as terrorist attacks, and a confidence in the fact that the new security measures could thwart such attacks. Troublingly, that is not the case. While terrorists certainly could target a mega-event, the real likelihood of their doing so is often a complete unknown. Analyses of terrorist activity have repeatedly demonstrated the "spatial displacement" of attacks away from hardened targets to softer, less defended areas. This has been particularly true for recent manifestations of violent Jihadi extremism (Coaffee 2009; Libicki et al., 2007). Hence, beyond a minimal threshold of basic security measures, elaborate and intense additions to the security presence at mega-events typically does little to reduce the prospect that terrorists might attack at the periphery of an event. Given the global media attention focused on such events, such peripheral attacks can still serve the terrorist's purposes. In light of the unknown probability of such threats, and the generally recognized inability of even the most elaborate security measures to eliminate the prospect of a terrorist incident initiated by a small group of dedicated attackers, critical scholars must search more widely to try and understand the massive financial and symbolic investments in mega-event security. Contributors to this book explore the various aspects of this issue.

Concepts and trends

This collection is motivated by two broad and contested concepts: surveillance and security. Both concepts have generated huge and diverse scholarly literatures. Both tend to mean one thing in academic circles, and another in popular and media discourse. That said, the two phenomena are intimately linked. While a host of technological, institutional, and human factors are involved in providing security, surveillance is now a routine component of such initiatives.

The analysis of security at mega-events occurs within a context of deepening concern about the levels and nature of surveillance, which is rendering our lives, activities, movements and behaviors more transparent than ever before to an increasing number of individuals and organizations. In the popular mind, the word "surveillance" tends to imply video monitoring or espionage. We regard it as a far broader phenomenon that structures relations between individuals and organizations, and indeed between individuals and individuals. As David Lyon notes, it comprises "any collection and processing of personal data, whether identifiable or not, for the purposes of influencing or managing those whose data have been garnered" (2001: 2).

Several overlapping trends contribute to this condition. The first is one of routinization. As citizens engage in the typical activities of modern societies they leave traces of data behind them – booking a hotel, surfing the Internet, using a credit card, booking an airline ticket, or attending an international sporting event. A huge proportion of surveillance is now about monitoring everyday life (Lyon 2001): who we are, what we are doing and where we are doing it. Everybody surrenders his or her personal information in exchange for a range of perceived benefits. Sometimes that surrender is voluntary and transparent; at others it is more secretive and coercive. The upshot is that one does not have to be a "suspect" anymore to be a subject of surveillance.

Contemporary surveillance cannot be accurately captured either by the metaphors offered by Orwell's "Big Brother" or that of Foucault's "panopticon" (Haggerty 2006). It is now more fragmented and dispersed throughout different computer networks within government, outside government, and within the grey areas in between. The networked information environment has created a more diffuse and elusive pattern of surveillance, which has eroded traditional institutional and functional distinctions, and disrupted existing hierarchies of visibility. Flows of personal data now percolate through a variety of more porous, and less discrete systems: "The resultant 'surveillant assemblage' operates by abstracting human bodies from their territorial settings, and separating them into a series of discrete flows. These flows are then reassembled in different locations as discrete and virtual 'data doubles' " (Haggerty and Ericson 2000: 605). It is disconnected and disembodied (Lyon 2001). And while these systems make people transparent to diverse audiences, it is becoming increasingly difficult to render the systems themselves transparent in hopes of curtailing or regulating them.

These trends also go hand in hand with revolutionary advances in information technology. Every year information processing becomes more powerful, and has now reached a critical threshold where it is often cheaper and more efficient to retain data than to delete it (Mayer-Schönberger 2009). Information technology is distributed and mobile, creating powerful incentives and capabilities for the surveillance of location and movement. It is miniaturized, and can be embedded in material objects, meaning that we are starting to witness an "Internet of things." And it is increasingly associated with advances in biometric identification, where the technology is not simply "about" the body, but is increasingly part "of" the body (van de Ploeg 2003). These trends are apparent during the experimental laboratory of the mega-event, as officials yield to multiple pressures to apply the latest and the most sophisticated technologies, whether that is RFIDs on tickets, facial recognition video-surveillance, or biometric identification at borders.

Surveillance is also increasingly global. With the rise of the transnational corporation employing private and public international networks, as well as the advent of "cloud computing," surveillance becomes a process that might originate from almost anywhere. Hence the ability of any one jurisdiction or organization to control surveillance is inescapably linked with the actions of public and private organizations that operate outside its borders. The development of the range of different processes that are embraced by the concept "globalization" gives the protection of personal information a character that is far wider than can be contained within the borders of single states.

These trends also go hand in hand with more general pressures towards securitization, where more issues, problems, behaviors and phenomena are being defined in "security" terms. If something is "securitized" it does not necessarily mean that the subject is materially important or critical for survival. Rather, it means that someone in power has constructed that "something" as a security problem. The ability to effectively securitize a given issue is then highly dependent on the status of a given actor. If a subject is successfully securitized, it is then possible to legitimize extraordinary means to solve that problem with the concomitant result that it can then be considered an inappropriate topic for critical debate or opposition (see Buzan et al. 1998). Mega-events are clear examples of such processes. As Pete Fussey and Jon Coaffee remind us in their contribution to this volume, Olympic security threats are not exogenously defined, but are selectively and socially constructed.

Cumulatively, such processes and technologies have ushered in a new era of transparency. This new transparency makes visible the identities of individuals, workings of institutions and flows of information in entirely new ways. Surveillance, the social process underlying the new transparency, is rapidly becoming the dominant organizing practice of our late modern world. Given growing computer-dependence and reliance on personal data collection and processing by a variety of institutions, and heightened public concern about security, surveillance is now experienced as an everyday reality. Never before has so much personal information been circulating about every aspect of our lives. As individuals are increasingly enmeshed in circuits of visibility it becomes harder to remain anonymous (Ratliff 2009), producing a progressive "disappearance of disappearance" (Haggerty and Ericson 2000: 619).

Finally, the particular dynamics of the terrorist attacks of September 11, 2001 (9/11) have clearly given a boost to securitization and surveillance and the legal mechanisms that have legitimated such practices (Lyon 2003). Whether or not 9/11 materially changed the balance between security and civil liberties, or simply acted as a catalyst for trends that were already under way, is a question the answer to which will depend on the context. However, the "moral panic" surrounding 9/11 and subsequent terrorist incidents clearly has had an impact on staging mega-events, including those discussed here.

What has shifted since 2001 has been the scale, technological innovation and centrality of surveillance strategies to overall Olympic security planning. Security has become an integral part of the Olympic ritual. The 9/11 attacks destroyed or seriously re-calibrated existing notions of security "proportionality" – the notion that the level of security should be proportional to the risk of untoward event. Given that the risks that are most on the minds of security planners are high-consequence but low-probability attacks, security officials have found themselves in an environment unconstrained by established notions of the empirical likelihood of attacks. The upshot has been that the factors that became most prominent in limiting the level of security were their costs and the degree of security presence that corporate sponsors were willing to stomach – both of which were ratcheted upwards. Security preparations have now reached the point where they are more appropriate for fighting a conventional war than protecting a soccer match from what most security analysts acknowledge would only be at most a handful of assailants that would be difficult to identify and thwart, even with the most elaborate security preparations.

In summary, the contemporary literature leaves us with the overwhelming message that the quantity and quality of monitoring have changed. It also strongly suggests that the traditional means of resisting these trends, framed in terms of protecting "privacy," are inadequate. "Privacy" and all that it entails is often regarded as too narrow, too based on liberal assumptions about subjectivity, too implicated in rights-based theory and discourse, insufficiently sensitive to the social sorting and discriminatory aspects of surveillance, culturally relative, overly embroiled in spatial metaphors about "invasion" and "intrusion," and ultimately practically ineffective (Bennett 2011 forthcoming). As a concept, and as a way to frame the various global challenges encountered within "surveillance societies," some contend that it is profoundly inadequate, and can never be the "antidote to surveillance" (Stalder 2002).

To be sure, all the various structural conditions identified within the surveillance literature come together before, during and after the events discussed in this volume: the pervasive logic of securitization, the obsession with high technology, the "protectionist reflex" (Beck 1998) manifested in pre-emptive risk assessments, the permeation of public authority to private interests, the passivity of most citizens, and the weakness of institutional resistance articulated through the lens and procedures of privacy protection. However, there is also an apparent sense that contemporary accounts of surveillance do not tell the whole story with respect to the "mega-event." There is clearly something extraordinary about these occasions, which requires further explanation and analysis. Other factors act as important catalysts.

The catalysts

Mega-events have become significant moments in the highly competitive efforts at "place branding" or "destination marketing" in which every nation, city, town and locality now feels compelled to engage. This competition for tourists, visitors, investors and residents is now integral to the dynamics of the globalized economy.

When done well, such efforts are expected to produce far more than a superficial logo, but a continuous, strategic and coordinated distillation of what residents are believed to consider the core characteristics of where they live. The place brand should be memorable and enduring in the public consciousness and should ideally shape and encapsulate further social, economic and cultural development. Place branding thus supports a multi-million dollar consulting and public relations industry (Ward and Gold 1994).

The mega-event, and particularly the sporting mega-event, is perhaps the ultimate opportunity for place branding and marketing (Gold and Gold 2007). It is also the signal occasion for the multitude of marketing consultants to coordinate, advise and promote. Thus, British Columbia was hailed as "The Best Place on Earth" in plenty of time for the 2010 Winter Olympics. The organizers for London 2012 contend that: "We need a powerful brand to help us achieve our ambition. A brand that combines the power of the Olympic rings and the city of London together. The number *2012* is our brand. It is universal and understandable worldwide. Our emblem is simple, distinct, bold and buzzing with energy... Its form is inclusive yet consistent and has incredible flexibility to encourage access and participation... It feels young in spirit. Full of confidence, certainty and opportunity" (London 2012).

Hosting mega-events also needs to be understood on a larger geopolitical scale. For some countries, South Africa being the primary example, the ability to successfully host a major event also becomes a badge of belonging, an affirmation of "first class citizenship in international society," as Daniel Bernhard and Aaron Martin argue. David Murakami Wood and Kiyoshi Abe note that similar dynamics have operated in Japan, which has continually used the staging of mega-events as a way to reinforce its status as a global power.

A corollary to the excessive hype and positive imagery is an overwhelming motivation that "nothing should go wrong" that might taint the carefully constructed brand. Of course, the "nothing" might range from a full-blown terrorist incident such as those experienced at the Olympic Games in Munich in 1972 or Atlanta in 1996, to less publicized threats from a range of right- and left-wing extremists, ethnic separatist movements and single issue groups. There are also a number of less harmful, and more contested, incidents and/or unsavory images and practices that become constructed over time as security problems: domestic protest, critical signage, homelessness, drug addiction and trading, panhandling, squeegeeing, graffiti, public fighting, and unlicensed street vending. Hence, there is a large array of potentially damaging incidents, activities and conditions, over which organizing committees may, or may not, have much control, at least not without trampling upon important civil liberties. For instance, Adam Molnar and Laureen Snider report the heavy-handed efforts to quell protests by anti-2010 games activists before and during the Winter Olympic games in Vancouver, from crude attempts at harassment to constitutionally suspect by-laws about negative signage along major Olympic routes. Volker Eick notes that Olympic officials policed even the food brands sold at the fan zones during the FIFA World Cup in Germany 2006.

In some jurisdictions, at least those with more established democratic traditions, excessive surveillance and control, or at least the visibility of such measures, can detract from brand images of openness, tolerance, democracy, and respect for constitutional rights. Excessive security sometimes sits uneasily with branding messages that often emphasize inclusiveness and participation, such as that promoted in London 2012: "Our Games will be for everyone."

Promoting a positive brand is not only, of course, of local concern. International organizing committees, such as the IOC and FIFA, place considerable, and perhaps increasing, pressure upon cities and nations. The host governments must now make elaborate commitments about how they will promote the event, and about their security arrangements. Several papers in this collection highlight this external pressure. Most notably, Minas Samatas documents the concerted efforts by American interests, fuelled by suspicions that Athens simply was not prepared to host the 2004 Olympics, to purchase and install an enormously expensive surveillance system from SAIC-Siemens, with attendant cost-overruns and allegations of corporate corruption which are still being sorted out. For Eick, the World Cup itself is a "brand" promoted by FIFA as a cash machine for securing profits.

In the aftermath of 9/11 there has been an enormous growth in the security industrial complex which has targeted mega-events as a lucrative opportunity to sell advanced security products. Hence, the drivers are often not what is necessary in light of the objective risk – something that is hard to determine – but what is the latest and greatest. From security guards to fencing to high-tech sensors and monitoring devices, all these corporations see mega-events as an opportunity to secure lucrative contracts and to showcase their products for future events. More generally, as Philip Boyle notes, transnational security companies often see their role in securitizing an event as a gateway to broader influence within domestic markets, and especially within the developing world.

There is therefore a powerful confluence of institutional motivations, all of which point not only to greater securitization of mega-events and the deployment of an increasing range of surveillance measures, but also to the central role of security in overall event planning. The city or nation wants to promote its brand. The security consultants and providers want to show off their wares. Local officials want to secure infrastructure that will endure after the event is over. International organizing committees want everything to go off without hitch. Corporate sponsors want return on their investment in advertising expenditures. The general public, both local and visiting, is willing to endure extraordinary measures. In the face of such a powerful confluence of actors and motivations, meaningful local opposition to spectacular security measures is difficult, if not futile.

The legacies

Sponsors for mega-events recognize the political delicacy of requiring host countries to pick up the massive tab now associated with such events. Consequently, they have fixated on the idea of "legacies," to argue for why such events are desirable. Such legacies are diverse and operate on several scales simultaneously – economic, symbolic, infrastructural, security and so on. While we now know that many of the most hyped legacies pertaining to jobs and economic development often fail to materialize or produce lasting effects, this has not mitigated the incessant reference to positive legacies by event organizers.

Talk of legacies pervades the justifications for the massive security investment for mega-events, where officials often note that the security technologies and expertise will endure in the host country long after the event has concluded. The chapters in this volume report a variety of such legacies that operate at a technological, informational, legal, geographic, and cultural level. Not everyone is necessarily enamored with the prospect that the security initiatives for an exceptional event will endure and in some contexts critical efforts have been directed towards ensuring that excessive security is dismantled at the completion of the event.

Technologies do, however, tend to endure, and they get used for purposes other than those for which they were originally developed and deployed. The idea that once technologies are set in motion they tend to follow their own course independent of human direction is a powerful theme in political and social thought (Winner 1977). Technologies cannot be understood outside their social and political context, and therefore can never be regarded as an independent force. They are always shaped in some way by the conscious and autonomous decisions of political agents, or by existent organizational norms or standard operating procedures. Nevertheless, there is often a compelling process of "function creep" at work, where the continued and strategic potential of a technology becomes more and more obvious over time to those who use it.

In this collection, the debate about technological endurance or function creep has been engaged most notably with respect to decisions about video-surveillance. Each event discussed in this volume has been attended by extensive use of publicly installed cameras at event sites, as well as more widely. To be sure, there is plenty of evidence of "function creep." Equally, there is evidence that officials in some locations have consciously planned for new systems to be used for other purposes after the event is over. In Athens, cameras installed at the time of the Games are still used for crime prevention and traffic control. In Geneva, a comprehensive video-surveillance system with the telling title of "Cyclops" was installed for the 2008 European football championships. This system, as Francisco Klauser reports, was introduced with the express purpose of monitoring the city after the event, and enabling data protection legislation was passed subsequently to legitimate its introduction. Fear of European, and particularly English, hooliganism during the 2002 World Cup in Japan prompted authorities to install video-surveillance in cities that were not even hosts to the matches and in other areas of others, such as red-light districts, that were predicted destinations of the visiting fans, according to David Murakami Wood and Kiyoshi Abe. More sophisticated systems with face-recognition capability were also introduced in Tokyo's Narita and Osaka's Kansai airports.

It is nonetheless difficult to assign clear cause and effect. Some of these schemes might have accrued anyway as a result of the more typical pressures documented in the literature on video-surveillance, not only pervasive fear of crime stoked by the popular media, but also liberalized urban planning policy and the ready availability of state funding (e.g. Norris and Armstrong 1999). It is obvious, however, that the mega-event acts as a catalyst, and an opportunity for local and national law enforcement to install systems that would be more difficult to justify in more "normal" conditions. For example, local officials in Turin used the opportunity of the Winter Olympics in 2006 to develop a completely new metro system, complete with automated, driverless trains that conduct continuous video-surveillance monitoring in real-time.

There is often heated local debate about such issues, but rarely a reversal. One possible exception to this trend occurred in Vancouver at the 2010 Winter Olympics. Around 1,000 cameras were installed by the RCMP and the City of Vancouver for the duration of the Games but the British Columbia Information and Privacy Commissioner has insisted that the surveillance system introduced to downtown areas should not be allowed to persist, at least not without appropriate public debate and approval from his office (CBC News, January 18, 2010). The city has now deactivated most of the cameras. However, there is still some hope by city officials that a hundred or so might be re-installed and that the monitoring center created for the Games will continue to operate (Stelter 2010).

Technological legacies are not confined to video-surveillance. A range of more opaque and interconnected databases support the surveillance operations associated with mega-events. Eick reports on the controversial German data bank of violent sports offenders (*Gewalttäter Sport*), which was introduced in 2000 and used during the 2006 World Cup. It still operates, notwithstanding the fact that it was declared unlawful by the Higher Administrative Court.

These examples accentuate the next form of legacy – informational. Staging mega-events provokes the capture of a variety of forms of personal information on a range of individuals, whether suspects or not – fans, athletes, employees, attendees, visitors to the host city and so on. This legacy is more difficult to substantiate and measure than the more visible technological manifestation of surveillance cameras. Nevertheless, this collection presents evidence that databases also endure, as does the personal information that they contain.

Databases entail matching and sharing, between a number of public and private stakeholders. Stefanie Baasch reports on the extensive sharing of data before, during and after the 2006 World Cup between police and armed forces at both federal and state levels, in possible contravention of German constitutional principles and its strict data protection laws. Samatas describes the C4I system, standing for "command, control, communications, and integration" that was supposed to operate during the Athens Olympics in 2004 and which involved a huge security infrastructure of cameras and databases. Giovanni Pisapia and Chiara Fonio report a similar integrated security network in Turin for the Winter Olympics in 2006. The system there even included an enhanced health surveillance network aimed at detecting and reporting incidents of specific infectious diseases to a central database.

One of the most intrusive, and probably most publicly debated, sets of measures associated with sporting events involves testing athletes for performance-enhancing drugs (Sluggett, forthcoming). International sporting bodies, such as the IOC, now have elaborate sets of protocols designed to prevent drug use before, during and after sporting events. Testing procedures have become more intrusive, often abandoning principles of informed consent and national constitutional guarantees against unreasonable searches and seizures (Park 2005). Random and mandatory testing, often unlawful without probable cause, is now the practice before any Olympic Games.

The results of these tests are held in databases coordinated through the World Anti-Doping Agency, an international body responsible for implementing the World Anti-Doping Code, and which is supposed to adhere to a voluntary set of privacy guidelines (WADA 2009). WADA then acts as a central clearinghouse for Doping Control Testing results for those athletes who have been included in their "testing pool." To facilitate coordination, each anti-doping organization is supposed to report all in-competition and out-of-competition tests on such athletes to the WADA clearinghouse as soon as possible after such tests have been conducted. This information is then made accessible "to the Athlete, the Athlete's National Federation, National Olympic Committee or, National Anti-Doping Organization, International Federation, and the International Olympic Committee or International Paralympic Committee" (WADA 2009: 3). Thus, personal biological information about athletes also endures in institutions and databases.

Knowledge endures, and particularly knowledge about security measures and their supposed effects. Boyle traces the various procedures through which this "know-how" about security circulates around the globe, and from event to event. The network now includes state and local law enforcement bodies, public safety and intelligence agencies, international sporting federations, international governance organizations, and security consultancy and technology firms along with a host of mediating actors including event management and logistics firms, industry associations, and public policy think tanks. This is, according to Molnar and Snider, part of a "mega-event security development nexus" which transcends public/private and civilian/military boundaries.

The knowledge transmitted through this network is not depoliticized, rational and instrumental. It is deeply implicated in a range of assumptions about security and risk that circulates around a densely networked and international "epistemic community," which in turn shares lessons about command strategies and structures, communications, and operational planning with different hosting administrations and officials, and with the specialized institutional security units that now seem a permanent feature of event planning. Moreover, the construction of what constitutes a "risk" tends to expand over time, as consultants and officials try to "think the unthinkable" about what possible disruptive event might occur. Fussey and Coaffee demonstrate this "shifting topography of risk" at mega-events over time. Molnar and Snider also provide a glimpse into the expansive understanding of risk that operated at Vancouver's Olympics, where a Joint Information Group addressed six main risk factors: (1) Financial Security/Organized Crime, (2) Terrorism, (3) Public Order, (4) Emotionally Disturbed Persons (EDPs), (5) Information Technology Security, and (6) Public Health.

Legal measures endure. Some of the chapters note examples where new legal provisions are introduced to manage a particular event, which then turn out to be permanent in nature and effect. As Molnar and Snider report, a controversial signage by-law, which regulated the "number, type, form, appearance, and location of signs throughout the city in order to reduce visual clutter" was passed by the Vancouver city council in advance of the 2010 Winter Games. Signs celebrating the Games and that enhanced the festive atmosphere, and approved by Council, or those posted by sponsors would be permitted. After a constitutional challenge under Canada's Charter of Rights and Freedoms, the city amended its law. While this by-law was only a temporary measure and ceased effect in March 2010, it did set a precedent for future events.

Eick reports that the government of Hesse amended certain laws allowing the extended use of video-surveillance systems, the extensions of custody for suspects from 48 hours to six days, automated number-plate recognition, and wiretapping. In Switzerland, a new data protection law was passed to permit more extensive use of video-surveillance during the 2008 European Championships. Pisapia and Fonio attribute the rushed passage of new anti-terrorism measures in Italy, not only to the London 7/7 bombings but also to the need to protect the Winter Olympics in Turin in 2006.

Spatial changes also endure, affecting the geography of cities. Mega-events can have enormous structural implications for host cities and nations. Probably more noticeable during the Olympics rather than the FIFA World Cup, where hosting responsibilities are shared between several cities, spatial legacies nevertheless affect the long-term potential for urban surveillance, crime control and the marginalization of certain groups. These events act as crucial occasions for experiments in crowd control, as well as the construction of city spaces, which produce economic value for investors and sponsors. The most notable instance of the latter was the "Fan Zones" organized for the 2006 World Cup in Germany. Pisapia and Fonio describe the various rings of security ("controlled, soft and hard") that structured the Winter Olympics in Turin, and which continue to have legacies in that city.

Some cities are, of course, more amenable to reconstruction and the creation of "spaces of exception" than others. In Sydney, for instance, a case not included in this volume, a totally new community was built 10 miles north of the downtown core to host most of the events. Much of the motivation was security-related, permitting a security perimeter surrounding the entire park, overseen by the Olympic Security Command Center (Sydney Olympic Park 2000). We are also seeing extraordinary transformations in London in preparation for the 2012 Olympics. Fussey and Coaffee report how the future Olympic sites in the East

End of London are being reconstructed into a secure "Olympic Ring of Steel" complete with electrified perimeter fencing, sophisticated video-surveillance systems built on the area's previous experiments with face recognition, screening points using advanced biometric checks and secure traffic free zones. These measures build on London's reputation as being one of the most monitored cities in the world, and also reflect its tradition of delineating "separated spaces" to tackle crime, terrorism and other risks.

Attitudes about surveillance and security also endure. Although more intangible and difficult to measure, it is nevertheless probable that mega-event "states of exception" (Agamben 2005) produce enduring legacies in terms of a broader cultural acceptance of new security practices and social hierarchies. The dynamics of securitization point towards the wider legitimation of surveillance, regardless of whether or not "things go wrong." If no threat to security occurs, planners can boldly assert that their extensive measures have "worked." If a threat does materialize, it can be conveniently explained by unforeseen gaps in the infrastructure and used to ratchet up levels of security at the next event. Molnar and Snider remind us, therefore, that the argument about exceptionalism should not be overstated. Mega-events are also a reflection of the prevailing neo-liberal order and serve to intensify existing insecurities and inequalities.

The collective message of this volume suggests that the citizens of host cities and nations are willing to endure extraordinary security measures that they may not tolerate in more "normal" circumstances. Many of these measures operate away from a public consciousness understandably focused on the spectacle of the competition, rather than on the RFID tags on the tickets, the enhanced security at the airport, the vehicle checks or the overhead surveillance blimps. Critical commentary on excessive security is also in danger of being conflated with a general nay saying against hosting the event in the first place. The festivalization of the event generally entails overwhelming public support, a broad intolerance for criticism, and an unwillingness to think too critically about long-term costs, financial and otherwise. Supportive public opinion is often claimed with reference to questionable and false dichotomies expressed in the popular media, as well as in opinion surveys; "would you rather have your privacy or be protected from terrorism?" It is possible, therefore, that the long-term effects of the "security games" lie not in specific practices or laws, but in a general consensus that incremental extensions in surveillance are a price worth paying for a "successful games, without incident."

The variations

There are, therefore, compelling and common international trends at work. The structural conditions and the exceptional circumstances of the mega-event are clearly powerful forces and there are strong continuities across space and time. It would clearly be a mistake, however, to conclude that these trends operate in a deterministic and uniform manner. Security processes are mediated and perhaps constrained by a number of local conditions that affect how the "security games" are manifested on the ground. The pattern of securitization is, therefore, likely to vary according to various structural and cultural factors.

The structural configuration of states invariably influences how issues are defined, debated and resolved. In the same way that identity card development is influenced by the administrative organization of states (Bennett and Lyon 2008: 15–16), some configurations are more conducive to securitization and surveillance than others. For example, mega-events, especially those hosted in several cities, require continuous bureaucratic cooperation over sharing and processing existing personal information systems. That cooperation will be more forthcoming in societies in which horizontal linkages are institutionalized across the bureaucratic divide. The need for inter-agency cooperation is not only a matter of administrative culture, but also one of technical inter-operability. Agencies must want to share the personal data within their custody, but they must also be able to do so. The legacies of information system design as well as technical standards sometimes impose significant constraints on building the databases necessary to support the kinds of integrated security provisions manifest during these events.

A further factor relates to the centralization of state functions. There are, for example, some important institutional differences between the states studied here. The existence of federal structures, for instance, has a clear impact on the conduct of centralized security planning. Klauser emphasizes this point in his comparisons between Switzerland and Austria in their approaches to hosting Euro 2008. Swiss federalism increased both the complexity of internal security coordination for Euro 2008 in general, and the scope for diverging regional and local security solutions more specifically. It also provided many different channels for critical political engagement with the planning process for Euro 2008. Baasch reports that the possibility of military operations by the Bundeswehr (Federal Armed Forces) during the 2006 World Cup was prevented by prohibitions within the federal constitution. Other states are based on unitary structures and with more centralized allocations of power, which can override local objections and interests. We probably see the effects of centralization in the more coordinated security arrangements displayed in Athens in 2004 and London in 2012.

The extent of securitization might be more prevalent in states that are more permeable to private sector influence. Several papers draw attention to the "security-industrial complex" comprising a network of public and private sector actors. Some state institutions are more permeable to these influences than to others. The vulnerability of the Greek state certainly comes through loud and clear in Samatas' chapter on Athens 2004, with significant implications for national sovereignty.

A range of cultural factors may also influence the local responses to event securitization. We might potentially expect a greater tolerance for extensive forms of surveillance in cultures that have historically exhibited higher levels of trust toward state authority. It is always interesting to hypothesize that the balance between privacy, civil liberties and security within different societies will vary according to their cultural traditions. However, political culture is a problematic and slippery concept, as Murakami Wood and Abe remind us. The traditional and stereotypical picture of a deferential Japanese culture may have little to do with the intrinsic attitudes of Japanese citizens, and more with the effects over time of coordinated practices of surveillant control. And it is always difficult to separate the intrinsic and perennial orientations to authority, from the more temporary set of attitudes about specific events and practices. Where there has been a concerted opposition to extensive securitization, as for example in Vancouver 2010, this resistance is often conflated with a range of wider questions about the financial, environmental and social impact of hosting the event in the first place. Security is only one prominent theme in the broader politics of mega-events, which is, at the same time, grounded in specific local contexts and conflicts.

Civil society activism is also supplemented by the more precise and technical critique of the arrangements for the collection, storage, processing and distribution of personal data in the many information systems that support event securitization. In many societies, this advocacy role is institutionalized within a data protection authority, such as the federal and provincial Privacy Commissioners of Canada, the Information Commissioner in the UK, or the federal and state data protection authorities in Germany (Bennett and Raab 2006). Typically these agencies do not have the political and financial resources to affect system development, especially since most privacy laws contain broad exemptions for law enforcement. But it should also be remembered that the collection and processing of personal information during these events should abide by pre-existing privacy or data protection laws, and systems should then be constructed with due regard to statutory information privacy principles. And those commitments are generally made. The organizers for London 2012, for instance, have published a privacy policy and declared commitments to compliance with the UK Data Protection Act. This policy, however, tends to expose some of the limitations of framing issues concerning security and surveillance in privacy terms. The policy is simply silent on the larger structural and spatial implications of the surveillance infrastructure of the sort that Fussey and Coaffee discuss.

Occasionally data protection agencies have asserted their authority. German data protection authorities were reportedly concerned about accreditation, video-surveillance and RFID chips on tickets for the 2006 World Cup (Dix 2007), and there was also some civil society resistance to this use of RFID "spychips" (FoeBud 2006). Before Vancouver 2010, the Privacy Commissioner of Canada sponsored research, hosted a conference in Victoria and had meetings with the Vancouver Integrated Security Unit. They also posted a dedicated Factsheet on "Privacy and the 2010 Olympics" in which the Privacy Commissioner reminded everyone that "the duty of governments to provide for the security of citizens must, in democratic societies, be tempered by the values that underpin our way of life. That is why the right to privacy must be upheld, even during mega-events like the Olympic Games, where the threat to security is higher than usual" (Privacy Commissioner of Canada 2010). So the issue was on the agenda and present in public debate, but it is not clear the extent to which these efforts materially affected security arrangements.

More generally, the abilities of these agencies to exercise their powers and responsibilities are always easier when there is a generally supportive public opinion. In the case of mega-events, as we have argued, the balance of public attitudes tends to be on the security side of the equation, and extraordinarily tolerant of exceptional measures under the impression, and perhaps illusion, that they will be temporary.

Implications and structure

This collection should not be read as a more general indictment of the Olympic Games, the World Cup or any other major international event, for that matter. The overall evaluation of the social, economic, cultural and international costs and benefits of hosting such events involves a complex set of judgments and calculations that go far beyond the question of security (Gold and Gold 2007). Rather, our intention is to foster constructive and critical debate on what we regard as a serious and worrying trend, which may in the long run prove detrimental to the spirit of competition, trust and fellowship that is supposed to underpin international sport. The underlying trajectory would predict an increasing attention paid to security, skyrocketing costs and probably a further limitation on the kinds of nations and cities capable of acting as hosts.

Many academic edited collections are criticized for being amorphous and lacking a singular focus. In contrast, this volume is precisely concentrated on the security and surveillance dynamics of mega-events, with many of the papers reinforcing and building upon the insights of other contributors. At the same time, this focus should not be mistaken for a narrowness of vision or a lack of social and political significance of the processes analyzed here. Mega-events are now a vehicle for massive forms of change that operate globally and at diverse scales. The political, cultural and financial desirability of such happenings now means that rarely does a year go by without one or many mega-events taking place, which further reinforces the fact that we are witnessing the paradoxical normalization of what are presented as exceptional happenings. The upshot of this proliferation of mega-events is also that a collection such as this could not begin to cover every event. That said, the case studies presented here are illustrative of both the standardizing tendencies as well as some of the axes around which variation can operate at different locations.

The collection begins with a broader overview by Daniel Bernhard and Aaron Martin of the geopolitical context within which the modern Olympics are held. The Olympic difference is attributable to the fact that hosting the Games offers the unique opportunity to demonstrate modernity and to acquire that elite status of a nation that can "build, compete, transport, and secure." Because of the Olympic difference, there is a much higher premium on Olympic security than on other events of equal size and vulnerability.

Pete Fussey and Jon Coaffee then trace the historical trajectory of Olympic securitization and demonstrate how the shift toward a coordinated "total security model" has intensified post-9/11. There are important continuities in the social construction of risk, which culminates in the planning for the future 2012 Games in London. The authors describe the particular security and surveillance configurations for the London Olympics. Although these measures reflect the continuing process of Olympic securitization, they are also layered upon a highly developed institutional experience in creating "technologically patrolled splintered spaces as a foil to terrorism." In many respects, the English capital *already* exhibits many of the characteristics apparent in the security practices present in prior mega-events.

Minas Samatas then explores the first post-9/11 Olympic spectacle, the Athens Olympics of 2004. He presents a critical report on the "super-panoptic" systems planned, and generally never implemented, at the Athens Olympics. Extraordinary pressure from the IOC, in league with American and other Western governments and powerful transnational corporations, imposed a highly coordinated surveillance system throughout Athens, in the belief that the Greek Games were particularly vulnerable to terrorist attack. Samatas analyzes the implications for Greek sovereignty, for the Greek economy and for the civil liberties of the Greek people.

In its desire for acceptability as a global power, Japan has a relatively long history of projecting its image to the outside world through the lens of the mega-event. Beginning with the Tokyo Olympics, through the Winter Olympics in Hokkaido and culminating in the 2002 World Cup, co-hosted with Korea, David Murakami Wood and Kiyoshi Abe analyze, not only the security arrangements at these events, but also the changing relations of Japan with the outside world. Japan has a long history of framing its national and cultural identities in terms of, and against, the conception of various "Others" that must be excluded or monitored to ensure the social order. International sporting events are a crucible for these various governmental and social anxieties. They reflect the evolving eras in post-war Japanese development.

We then present two papers devoted to the 2006 World Cup in Germany. Volker Eick offers a critical political economy analysis of FIFA and the 2006 World Cup, which saw the largest display of domestic security strength in Germany since 1945. The football venues as well as the "fan miles" in downtown areas were converted into high-security zones with access limited to registered persons, and overseen by sophisticated surveillance, information, and communication technologies. Eick explains this extensive "security assemblage" in terms of FIFA's neoliberal agenda to promote its brand (The World Cup) and enhance its profits. Stefanie Baasch investigates the impact of the 2006 World Cup on urban development. She demonstrates how the construction of particular threats and risks in the German media are used to legitimate characteristic security policies and practices, both during the event and afterwards.

Francisco Klauser studies the 2008 European Football Championships co-hosted in Austria and Switzerland. Different national and urban comparisons allow him to reassert the role of local agency, motivation and expertise in the security governance of major sporting events. The profile of mega-event securitization as an externally imposed operation should not be exaggerated. There are a number of structural features, particularly manifested in Switzerland, a federal country that is not a member of the EU, which present a degree of local autonomy and maneuverability. Giovanni Pisapia and Chiara Fonio also stress the importance of local conditions in their chapter. They offer a detailed analysis of the Integrated Security System developed for the 2006 Winter Olympics in Turin, the various local administrative arrangements and the legacies, especially in communications and transportation infrastructures.

The collection concludes with two papers on Vancouver 2010. Adam Molnar and Laureen Snider, unlike other authors, do not see the Games merely as a performative ritual, or as a profit-generating zone of consumption, but as a broader cultural political economic model of development. The accelerated and temporary investment in mega-events, often at the expense of other social goods, can have long-term marginalizing effects on vulnerable populations which often find articulation in securitization strategies. Finally Philip Boyle analyzes how the international networks operated to transfer "know how" about security to the Vancouver 2010 Games. He describes the institutional arrangements and the international, national and local actors. He also demonstrates how particular assumptions about security and risk motivate the discourse, the actions and the legacies.

In a time of limited political vision and deep public cynicism, mega-events are presented to the public as a panacea; a spectacle capable of uniting the nation and being all things to all people. They are undeniably vehicles for change, introducing transformations at a host of different social, cultural, economic and political levels that would otherwise be unlikely to occur. In recent years event security has become one of the most high-profile domains of change, producing a myriad of different sorts of legacies. While officials present the new security technologies and surveillance devices used at these events as being desirable and even inevitable, this book takes up the long overdue task of critically interrogating what is being secured, why and for whose benefit.

References

Agamben, G. 2005. *State of Exception*. Chicago: University of Chicago Press.
Beck, U. 1998. *World Risk Society*. Cambridge: Polity.
Beijing 2008. "Beijing intensifies security preparations for Olympics," at URL: http://en.beijing2008.cn/news/dynamics/headlines/n214244540.shtml (accessed August 18, 2010).
Bennett, C. forthcoming 2011. In Defense of Privacy. *Surveillance & Society* 4 (4).
Bennett, C. and C. D. Raab. 2006. *The Governance of Privacy: Policy Instruments in Global Perspective*. Cambridge, MA: MIT Press.
Bennett, C. and D. Lyon (eds.) 2008. *Playing the Identity Card: Surveillance, Security and Identification in Global Perspective*. London: Routledge.

Boyle, P. and K. Haggerty. 2009. Spectacular security: Mega-events and the security complex. *International Political Sociology* 3 (3): 257–274.

Buzan, B., O. Wæver, and J. de Wilde. 1998. *Security: A New Framework for Analysis.* Boulder, CO: Lynne Rienner Publishers.

CBC News. 2010. Olympic Surveillance Cameras Causing Concern. January 18. URL: http://www.cbc.ca/canada/british-columbia/story/2010/01/18/bc-olympic-surveillance-cameras-robertson.html (accessed July 9, 2010).

Coaffee, J. 2009. Protecting the urban: The dangers of planning for terrorism. *Theory, Culture & Society* 26 (7–8): 343–55.

Dix, A. 2007. "From Germany's Football World Cup 2006 to the UK's Olympics 2012: Reconciling mass surveillance, anti-terrorism and protecting privacy." Presentation to conference on Privacy Laws and Business, June 3, 2007. URL: http://www.privacylaws.com/templates/Page.aspx?id=1084 (accessed July 14, 2010).

FoeBuD. 2006. *2006 Soccer World Cup.* URL: http://www.foebud.org/rfid/en/world-cup (accessed, July 14, 2010).

Gold J. R. and M. M. Gold (eds). 2007. *Olympic Cities: City Agendas. Planning and the World's Games, 1896–2012.* London: Routledge.

Haggerty, K. D. 2006. "Tear down the walls: On demolishing the panopticon." In D. Lyon (ed.) *Theorizing Surveillance: The Panopticon and Beyond*, 23–45. Cullompton: Willan.

Haggerty, K. D. and R. V. Ericson. 2000. The surveillant assemblage. *British Journal of Sociology* 51 (4): 605–622.

International Marketing Council of South Africa. 2010. Brand South Africa. URL: http://www.brandsouthafrica.com/what-we-do.html (accessed July 9, 2009).

Libicki, M., P. Chalk and M. Sisson. 2007. *Exploring Terrorist Targeting Preferences.* Santa Monica: RAND Corporation.

London 2012. 2010. Our Brand. URL: http://www.london2012.com/about-us/our-brand/index.php (accessed July 9, 2010).

Lyon, D. 2001. *Surveillance Society: Monitoring Everyday Life.* Buckingham: Open University Press.

—— 2003. *Surveillance after September 11.* Cambridge: Polity Press.

Mayer-Schönberger, V. 2009. *The Virtue of Forgetting in the Digital Age.* Princeton: Princeton University Press.

Norris C. and G. Armstrong. 1999. *The Maximum Surveillance Society: The Rise of CCTV.* Oxford: Berg.

Park, Jin-Kyung. 2005. Governing doped bodies: The World Anti-Doping Agency and the global culture of surveillance. *Cultural Sudies – Critical Methodologies* 5 (2): 174–188.

Privacy Commissioner of Canada. 2010. *Privacy and the 2010 Olympics.* URL: http://www.priv.gc.ca/fs-fi/02_05_d_42_ol_e.cfm (accessed August 17, 2010).

Ratliff, Evan. 2009. Gone forever: What does it take to really disappear? *Wired* 17: 9.

Sluggett, B. (forthcoming). "Sport's Doping Game: Surveillance in the Biotech Age." *Sociology of Sport Journal.*

South Africa, FIFA. 2010. The Government's Promise. URL: http://www.sa2010.gov.za/safety-and-security (accessed June 23, 2010).

Stalder, F. 2002. Privacy is not the Antidote to Surveillance. *Surveillance & Society* 1 (1): 120–124.

Stelter, L. 2010. The Olympics: A Look Back. *Security Director News*, June 22. URL: http://www.securitydirectornews.com/?p=article&id=sd201006MH1pqJ (accessed July 13, 2010).

Sydney Olympic Park. 2000. Sydney 2000 Olympic Games History. URL: http://www.sydneyolympicpark.com.au/education_and_learning/history/olympic_history (accessed July 10, 2010).

van de Ploeg, I. 2003. "Biometrics and the Body as Information: Normative Issues of the Socio-technical coding of the Body." In D. Lyon (ed.) *Surveillance as Social Sorting*, 57–73. London: Routledge.

Vancouver Statement on the 2010 Winter Olympics. November 25, 2009. URL: http://www.sscqueens.org/Vancouver_Statement (accessed August 20, 2010).

Ward, S. V. and J. R. Gold. 1994. *Place Promotion: The Use of Publicity and Marketing to Sell Towns and Regions*. London: Wiley.

Winner, L. 1977. *Autonomous Technology: Technics-out-of-Control as a Theme in Political Thought*. Cambridge, MA: MIT Press.

World Anti-Doping Agency (WADA). June 2009. *Protection of Privacy and Personal Information*. URL: http://www.wada-ama.org/Documents/World_Anti-Doping_Program/WADP-IS-PPPI/WADA_IS_PPPI_2009_EN.pdf (accessed July 9, 2010).

Chapter 1

Rethinking security at the Olympics

Daniel Bernhard and Aaron K. Martin

The Olympics have become an object of security exceptionalism, wherein perceptions of danger are greater than normal and so the normal means of public security provision are greatly expanded. Because the Games provide a unique opportunity for countries to show their best side to the world and for cities to climb the global hierarchy of importance (Degen 2004; Short et al. 2000; Shoval 2002), most agree that security should equally rise to the occasion. In contrast to the self-evidence with which the urgencies of Olympic threats are normally presented, we seek to denaturalize the claim to exceptionalism and highlight some of the underlying social valuations that contribute to the Olympic state of alarm. Several contributors to this volume address the implications of specific security behaviours. We offer a conceptual framework within which the security mindset can be understood and debated.

We also offer an analytical hypothesis to describe some underlying motivations for the sense of emergency surrounding the Games. We attribute security policies to the macro-societal process through which the Olympics acquire their value, and through that value, their "endangered" status. The disparity between an Olympic security event and normal security measures deployed against threats to equally vulnerable assemblies of people connotes an "Olympic difference": the social designation of the Olympic Games as a special event that transcends the material parity of that event with other human security vulnerabilities. We view the Games as a *sui generis* object of the claim to security: they provide the state with an opportunity to affirm its modernity and therefore they become *objets protégés* in their own right. Security at the Games is not only deployed to protect that affirmation, it is also a part of it.

We maintain that membership in the club of Olympic hosts resembles many other élite clubs of nations, including the nuclear club, insomuch as hosting the Olympics becomes a pathway to, or alternatively, an affirmation of, first-class global citizenship. We unpack some of these commonalities to further our argument that the urgency surrounding the Games relates more to their (still real, though largely) symbolic importance to the host nation than it does to concerns about the safety of spectators or infrastructure. We do not think that security planners neglect human security or the protection of infrastructure in their planning.

Rather, we wish to account for many measures that clearly exceed the exigencies of protection.

Above all, we argue that the departure from rationalized security provision at the Olympics, and the main cause for the state of security exceptionalism, is the value governments place on the opportunity to demonstrate their full modernity to the world. Olympic security measures put human resources, technological sophistication, wealth, and organizational acumen on display in the same way that constructing ostentatious stadia or advanced weapons do. A significant effect of security at the Games is to reinforce this message by demonstrating not only that the Games can be held safely, but that safety can be achieved in the most modern, high-tech, and opulent way. Students of nuclear politics will be familiar with these status markers, and below we further explore that comparison. Those who are capable of hosting the Games must be able to build, compete, transport, and secure—none can be separated from the others and all contribute to the fundamental affirmation of élite status that is part and parcel of hosting the Games.

The Olympic mindset

There is an entrenched belief that the size of the Olympic platform is directly proportional to terrorists' desire to attack it. Even before September 11, 2001 (hereafter 9/11), fear that big events are likely to become "big killing events" was pervasive in the popular mind (Tulloch 2000: 230). Most scholarship on Olympic security takes the state of Olympic exception for granted. These works mention the post-9/11 "climate of insecurity" (Yu et al. 2009: 392) that affects security planning around the world in the most general sense (Boyle and Haggerty 2009: 257; Johnson 2006: 2; Voulgarakis 2005), and the increased visibility that an attack on the Olympics would bestow upon terrorists (Price 2008) to conclude that the Games are at a higher-than-normal risk of attack, meriting a harder-than-normal security posture. Even critics of the "security and surveillance industrial complex" make this assumption: "Especially in the post-9/11 world, the Olympic Games—as the most important global sports event and mega-media show—naturally cause very high security concerns" (Samatas 2007: 221). That governments place great symbolic importance on the Games—China's Games are the most potent example of this (Xiaobo 2008; Xu 2006)—only increases terrorist temptation, they say. These scholars may criticise the implications of Olympic security provision—outsourcing, corporate gouging, overkill—but not the motivations behind it. They suggest that a deviation from normal is necessary, though many stress the danger that a disproportionately strong security posture may not be extinguished with the torch (Head 2000). Even opponents of the Games agree that the Olympics need to be highly secured; the costs and effects of security operations underwrite their opposition (No 2010 2009).

In spite of the well-known fact that large killings are often sensational for their banality, perceptions of an Olympic difference are unchanged. Nobody can credibly say the Madrid, Mumbai, London, Bali, or Moscow bombings lacked visibility, media attention, or symbolic impact. These attacks preyed on the banal vulnerabilities of everyday life, not the spectacular exceptions of a mega-event. This is not to say that the attackers could not have attacked a more high-profile target, but rather that the impact of their actions was exploded by the constant fear they inspired in people going about otherwise simple activities. By this logic, the constructed importance of the Games would make them less of a target, not more of one.

We know this, but still there is something about the Olympics that radically alters society's representations of danger and the leeway we afford to governments to take extraordinary precautions to keep us safe. We argue that this general accord is the result of a securitization (Wæver 1995; Buzan et al. 1998): a broad consensus that the Olympics expose host societies to temporary but existential dangers from credible threats, and that the Games deserve to be protected, even if such protection requires the state to transgress the restraints we normally demand of it.

This construction may be widespread but it is not natural. Furthermore, it is far from certain what the referent object of the Olympic securitization—that thing deemed threatened and in need of protection—is. Accusations of excess and paranoia at the Games, we argue, depend on limited conceptual understandings of security that fail to differentiate between various representations of danger. They fail to ask: "In this case, *what* thing of value is being protected by a given policy? *From what* or *from whom* is that valued object endangered, and *why should we care* if it is?" If we ask these questions, we find that what seems like security exaggeration is only excessive if the referent objects of the Olympic securitization are people and buildings alone. Could there be others?

Defining security

When people talk about security, it is mostly unclear exactly what they mean. In the Olympic usage, security seems to mean all of the military, police, intelligence, and surveillance resources deployed around the host city, with the aim of preventing acts of violence. But even in its most simple formulation, security is about more than protection. We make war on terror, but not on traffic deaths, for example, even though the latter are materially more dangerous (Statistics Canada 2008; National Highway Traffic Safety Administration 2008). Though the material consequences (and therefore risks) of automotive accidents are always greater than those of terrorism, automotive fatalities are not considered matters of national security. The state does expend considerable resources to maintain and police roads, to ensure vehicle safety, and to reduce incidents of drunken driving, but a certain degree of risk is accepted and normalised. This is the realm of normal politics—where risks are known and accepted—not the urgent domain of security (Wæver 1995).

Where most authors on Olympic security do not probe the central concept of security, we use the constructivist understandings of security developed by the

Copenhagen School of security studies (Buzan 2007; Buzan et al. 1998; Wæver 1995) to (re)theorize the motivations and effects of Olympic security operations. The Copenhagen perspective allows a wider definition of what can count as security, providing us with theoretical tools to examine, through security measures, how the Olympics acquire their societal value.

As Buzan (2007: 106) points out, people experience threatening information along a spectrum from complacency to paranoia. Fear and courage emerge from the unconscious mind and only have a loose connection to rational calculations of risk. Where this connection is slow to arrive, the security priorities of the state can work to shape public anxieties in their image. Security and insecurity are emotions emerging from the interplay between threats, corresponding vulnerabilities and, most importantly, the perception of existential danger that results from their convergence (cf. Weldes et al. 1999). Security is thus entirely relational (Weldes 1996) and cannot be provisioned in one or another quantity.

Buzan's constructivism opens pathways to a wider security agenda (cf. Ullman 1983), where representations of danger can be triggered by threats from/on various "sectors" (Buzan et al. 1998), military and non-military, that threaten many potential "referent objects" other than the state. When we speak about security then, we are speaking about something more than just the probability of death or harm, something more than the cameras and cordons authorities use to militarise urban areas for mega-events. We must look to perceived threats, representations of danger, the thing declared endangered, and the societal consensus that allows governments to take exceptional, emergency action.

Wæver (1995) outlines how constructed representations of insecurity are operationalized through the "securitization" process. Whereas others use "securitization" to mean militarization (for example, see Yu et al. 2009), Wæver endows the term with a very specific nominalist signification. Securitization begins with the speech act, whereby someone announces a threat and tries to persuade the public and other major stakeholders that the threat is real, that it poses an existential risk to society, and that government must pursue a special urgency to remedy the situation and must therefore be temporarily granted the power to exceed the rules by which it is normally constrained. If the speech act is successful, the state escalates the situation by temporarily mobilizing its full resources to deal with the urgent matter. If the matter is no longer deemed urgent, de-securitization occurs, and the issue returns to the normal realm of politics.

Having earned the moniker of "security" and the full-scale mobilization of the resources of the state that accompany it, the Olympics have undergone a process of securitization—a broad consensus of their value has been reached—which we seek to present here. The security tag distinguishes the Olympics from "the political," those risks like traffic accidents which are deemed acceptable elements of normal life. That the Games marshal the urgency of security prompts us to explore how such great subjective value could be attributed to them, even when the objective risks to the Games are not as grave as one might suspect.

More than enough—risk and response

Regardless of whether events of symbolic, critical violence have increased since then, a post-9/11 consciousness dominates the security policies of advanced nations (Enders and Sandler 2005). The general climate of terrorist anxiety that permeates the fabric of world politics always makes a prominent appearance in the literature on Olympic security, and in statements by Olympic security planners, as a baseline against which the Olympic escalation can be understood. For instance, Voulgarakis (2005: 1), the minister responsible for Olympic security in Athens, describes the impact of the post-9/11 security consciousness on his own planning, and links general fear to the perception of risk at the Games:

> Soon after Greece assumed the responsibility to host an event that has been synonymous with peaceful competition ... the ugly head of irrational violence made its appearance as we entered the twenty-first century .... Indeed, the tragic events of 9/11 have turned the page of a new chapter in the history of humanity, and Greece was among the first countries faced with the awesome task of staging the world's largest sporting event under the threat of potential violence.

In addition, Voulgarakis shows that although being awarded the Games is a great honour, not all hosts are universally trusted to adequately perform their security duties. Only successful Games can convince the sceptics and fully confirm the host's competence. In fact, Voulgarakis argues that (mostly American) scepticism was unfounded: *anybody* would have been unprepared because no state had adapted mega-event plans to consider a scenario like 9/11. Though he does so retrospectively, Voulgarakis discusses the nature of threats to the Olympics, their credibility and existential essence, and the need for Greece to take overwhelming and abnormal measures to protect the Games, which for obvious reasons that need not be explained, he says, deserved protecting.

Many forget that the post-9/11 security environment does not apply just to mega-events: it is a constant feature of everyday life. By comparison, a recent report by the Canadian Security Intelligence Service (2008) prioritises the threat from a "lone-wolf" attacker at the Games. It may be the case that with $1 billion CAD and 15,500 officers on the case, the lone wolf is the only conceivable scenario with a chance of success. It might weigh on the minds of security planners precisely because it cannot be decisively planned against. But the same is true of the lone wolf in non-Olympic times, and we still allow other high-profile gatherings and provocative events to take place. We still fly on planes. In purely material terms, the consequences of a successful lone-wolf attack—measured in lives lost and property destroyed—are exactly the same whether or not the venue is being used for the Olympics. Governments are very successful at protecting people and property in normal times, especially against lone-wolf threats, and at a much lower cost. For $181 million CAD, Canada screens 42 million travellers in

89 airports every year (Duguay 2009: 48; Canadian Air Transport Security Authority 2007: 30) and has succeeded in securing air travel from all forms of explosives, firearms, toothpastes, and tweezers for about $4.40 CAD per passenger. Airport screening is the basic model for ensuring that no weapons get into stadia, but at the Olympics, the same tests cost about $2,579 CAD per person. About 18,651 spectators at a Vancouver Canucks match at General Motors Place merit 11–13 police officers, 20–22 traffic police and 45–50 private security staff to ensure a safe event (Jones 2009), and these events have been kept safe. The same crowd in the same place at the Olympics is protected by a parade of security agencies at much higher costs.

Summer Games depart even further from security normalcy than do winter festivities. Volugarakis deployed over 49,000 public and private security personnel and volunteers to the Athens Games (Johnson 2005: 310; Samatas 2007: 229)—more than 70,000 for the opening ceremonies (Samatas 2007: 224). Athens launched a network of nearly 3,000 video surveillance cameras, three police helicopters, and a surveillance blimp, all connected to the C4I central information security system that could broadcast images and sound in real-time to five main operation centres, streaming surveillance and dataveillance information into a comprehensive, multi-agency interface serving thousands of security officers (Samatas 2007: 224–226). Though C4I was not completed in time for the Games, it was an ambitious, high-tech element of "the biggest security operation in peacetime Europe" (American Society for Industrial Security, quoted in Samatas 2007: 221). The security effort in Beijing was even more surreal.

And so, the question arises: *what for?* If security operations stand between threats and vulnerabilities to protect something that is deemed valuable, then what are these things? As Samatas (2007) explains with no small irony, the much-lauded C4I system was not delivered on time for the Olympics, but the Games still proceeded without incident. C4I malfunctioned completely, with no security fallout. And if the worry is about terrorism, as declared, many measures seem out of place. The Patriot missile, for example, is primarily used for missile defence—to defend against complex ballistic missile systems which are controlled exclusively by sovereign states at the present time. These weapons could potentially shoot down aircraft if need be, but there are many alternative systems which could perform this same task, in a much more discrete and inexpensive way (cf. Canadian Security Intelligence Service 2008; Voulgarakis 2005). Still, Patriot missile batteries were a visible component of the Athens security strategy. The size of the force deployed to the Beijing Olympics was larger than the combined forces occupying both Iraq and Afghanistan. Was China expecting an invasion?

Some see the extreme reaction as a natural outcome of the security psyche. De Goede (2008) argues that with its self-escalating logic of premeditating disaster, the fundamental characteristics of security planning become extremely unbalanced, causing vulnerabilities to appear even in the absence of threats against them. Critics are fixed on incongruities between threats, vulnerabilities, and state responses, and as we have seen, objective justifications for some of the more elaborate Olympic measures are lacking. Flashy security measures are either mismatched to the threats prioritized by the intelligence community, or far surpass the already successful security provisions of everyday life. Moving backwards from Buzan's assertion of security being relational, a key question emerges: in response to what credible threat are these escalations enacted? This is more than a counter-factual question. Rather, it raises the possibility that the process of material risk assessment goes beyond concerns for the safety of participants and infrastructure. Could it simply be that we are looking to the wrong place, and that these measures could correspond to a threat-vulnerability complex beyond the terrorist-civilian cliché?

The Games as a *sui generis* object of security concern

People and places, though important, may not be the sole referent objects of the Olympic securitization. To fully understand Olympic security, we must look to the "Olympic difference," a subjective process of collective valuation that places a much higher premium on Olympic security than it does on normal assemblies of materially equal size and often greater vulnerability. Where normal operating guidelines would prohibit government from rolling out the big guns to deter instances of burglary or other petty crimes, public accession to a belief in Olympic exceptionalism paves the way for an awe-inspiring demonstration of the state's capabilities. Though the state protects the vibrant Olympic celebration, the means deployed, as we have seen, go far beyond the imperative to protect. Instead of accusing the state of overkill, we propose a way to make sense of Olympic security by widening the agenda of what counts as security to propose that the Games have become the referent object of a securitization all their own. If the Games are viewed as a venue in which the host nation and host city can affirm themselves and their advancement to the world, Olympic security measures *do* correspond to certain sets of threats and vulnerabilities, which are otherwise unseen.

Obviously, it is impossible to conclusively identify security planners' main concerns. We offer no "smoking gun" empirical evidence in defence of our claim and we do not mean to suggest that Olympic security planners focus on the symbolic at the expense of public safety. Rather, we feel compelled to offer a suggestive alternative to the naturalized self-evidence with which Olympic security is commonly portrayed. What follows is an attempt to map the practicalities of Olympic security measures onto the stakeholder model and process approach of the Copenhagen School, which lead us to hypothesise about an unseen referent object which most commentators miss: the Games as an existentially-salient political device which states use to affirm their prowess.

Where threats and vulnerabilities fail to correspond materially, a second look at the specifics of Olympic security measures can tell us a lot about the Games as a stand-alone object of securitization. What emerges is a demonstration that is both spectacular and sensational. Air forces, flotillas, missile batteries, special forces units, security cordons, search without warrant, surveillance technologies galore – none of these is a stranger at the Olympics. This is not just the kind of security theatre designed to feign complete control over the uncontrollable (Beck 2002: 41) in hopes of reassuring an anxious public about the state's ability to guarantee safety (cf. Boyle and Haggerty 2009), for, as we have seen, these measures go far beyond even the most fanciful imaginings of danger. What emerges above and beyond the crystal palace is a demonstration of the state's wealth, technological advancement, and organizational acumen. It seems fair to judge security against the same criteria used by the IOC to determine whether a given country can build the stadia, accommodations, transport networks, and communications infrastructures necessary for hosting a successful Olympics. We have long understood that the stadium-monuments and cultural pageantry of the Games demonstrate hosts' modernity and advancement, and ultimately act as proof of their suitability to host the Games. Where many view security separately from other aspects of the Games—a big stick to protect a big prize—it can also be viewed as a co-development of the Olympic project. As with the other demonstrations, Olympic security depends on vast human resources, the most modern technology, substantial riches, and the highest degree of organizational expertise. In the following sections we review how these different factors have served to make visible the security profiles of recent Olympic hosts.

Boots on the ground

Beijing's security operations were by far the most stunning on record, in no small part due to a massive human deployment. Force estimates vary, but official statements from Beijing organizers listed 92,500 people as being involved in the direct security of the Games. That figure does not include an additional 100,000 regular soldiers and 290,000 civilian security volunteers. This "saturation security strategy" (Thompson 2008: 50) allowed authorities to erect a security cordon around the entire city, search incoming cars for about 20 minutes each, and set up checkpoints at major junctions throughout the city (Yu et al. 2009: 393).

In Vancouver, the Integrated Security Unit coordinated more than 15,500 police officers, soldiers, and private security contractors (Thomas 2010). The Vancouver security contingent swelled so far beyond original estimates that organisers were forced to hire three commercial cruise liners to billet them all.

Modern technology

We have already discussed the degree to which Greek authorities centred their security demonstration around high-visibility high technology. This included Patriot missile batteries borrowed from the USA, and, of course, the C4I interactive surveillance and information-sharing system. Evidently, the failure of C4I had no impact on safety, but it was a massive security failure to the degree that the flash of the Olympics could only be met by the low-tech thud of soldiers' boots.

Further, most of the high-tech security in Athens, especially military hardware, was borrowed from the USA or NATO (Brianas 2004; Lynch and Cuccia 2006: 54–58). Though Greek officials were in charge, the technological demonstration was lacklustre. It mirrored poor performance on stadium construction and the like and was a blemish on the reputation of Greece and the organizing committee.

Chinese authorities assigned air force units to their Games and deployed anti-aircraft missile batteries in the Olympic Park. China's commitment to make theirs the "high-tech Olympics" included installing approximately 265,000 new surveillance cameras covering more than 50 percent of Beijing, and even inserting RFID chips into tickets to provide security screeners with the bearer's name, address, e-mail address, telephone number, and passport information. China also worked with Interpol to screen airline passengers against a worldwide database of over 14 million stolen document records and an even larger registry of suspicious persons.

Though Canada's military equipment is aging, it was on full display for the 2010 Games. Air force pilots flew CF-18 Hornets over the Games area, in support of Aurora strategic surveillance aircraft, Buffalo search and rescue aircraft, Griffon tactical helicopters, Cormorant and Sea King search-and-rescue helicopters, and RCMP helicopters. At least two frigates, two coastal defence vessels and several smaller craft patrolled the harbour, conducting surveillance and interdiction operations (Green 2009). The Navy also provided unique services such as underwater explosive ordinance disposal (Kitchen 2009).

Wealth

It goes without saying that security costs consume increasingly large portions of the Olympic budget. Where $300 million USD was deemed obscenely large at Salt Lake City, Vancouver spent approximately $1 billion CAD on security. Of course, the Olympic demonstration is as much about opulence as it is anything else. No stadium is too magnificent, and no security measure is too expensive for the Games. To host an Olympics without Ares would be to miss the point, no matter how much he may charge.

Organizational acumen

The sheer number of people, agencies, machines, and dollars allocated to Olympic security requires an impressive display of organizational acumen. Keeping this smorgasbord in order is no small feat. At Salt Lake City, the Secret Service directed security personnel from upwards of 30 discrete agencies through a consolidated command structure. Since then, the IOC (2004) has made the (legal and administrative) ability to develop a consolidated command structure a requirement of the Olympic bidding process. Another major organizational task is the creation of security cordons. Disrupting the flow of complex, highly-used spaces to reorganize whole cities for security purposes is a monumental task (cf. Klauser 2009).

Olympus: the peak atop the mushroom cloud

Rather than separate security from the Games it is said to protect, we can instead understand security as a manifestation of the greater affirmation that hosting the Olympics can be ostensibly reduced to. In terms of their cost, labour requirements, opulence, technological sophistication, and bureaucratic co-ordination, big security and big building projects start to look quite similar. As with infrastructure, demonstrating to the world that the state *can* draw on the finest modern security resources is an affirmation of first-class global citizenship. To this end, the difference between normal security and Olympic security lies in the prestige of the Games and the opportunity they provide for the host nation to affirm itself domestically and internationally. When security is included in this demonstration, it paradoxically serves to defend itself, existing as it does at the very heart of the affirmative process.

With such a wealth of material, human, economic, and bureaucratic resources behind them, it should come as no surprise that hosting the Games remains the *de facto* privilege of precious few countries. Behind the G8 and the nine nuclear states, the 22 nations deemed capable of hosting the Olympics with due fanfare and competency come a close third. Though they would derive no material benefit from hosting the Games—the cost alone would probably cripple them—even poor cities in poor countries are lining up to join the list, which will not expand until 2016, when Rio de Janeiro hosts the Summer Games. Just before being selected, officials in Rio launched an international media campaign highlighting the international political benefits that a successful bid would bring. As Barrionuevo (2009) put it: "Winning the Olympics would be a transformational moment for Brazil, an affirmation of its rising global importance."

Hosting the Games remains desirable even though the "national interest," in material or rationally calculable terms, is not clearly served by doing so. Several economists have attempted to discern whether the Games have a positive economic impact on host cities (Coates and Humphreys 2003; Owen 2005; Preuss 2004). Some believe they do and others believe they do not, but the long list of debtor hosts at least indicates that hosting the Games is not guaranteed to be materially beneficial, especially when the state shoulders most of the non-redeemable costs like security while private investors finance more lucrative projects like property development. Still, enthusiasm is hardly waning. As with other super-élite clubs in which status considerations often trump material arguments, the Olympic affirmation seems to transcend concerns of cost and debt. Consequently, it is telling to compare the Olympic affirmation to another kind of affirmation that defies material rationale: nuclear proliferation.

Like the Olympics, the net benefits of nuclear proliferation are uncertain. Though deterrence is still a popular argument (Joseph and Reichart 1998), a substantial literature now argues that nuclear weapons actually detract from state security. This is especially true during development: "Given the fears that new nuclear capabilities among neighbours create, there will be a temptation to destroy them by pre-emptive strikes while they are still small and vulnerable. . . [Aspirant nuclear states] may have to pass through a dangerous 'valley of vulnerability' " (Nye 1986: 90; cf. Goldstein 2003). Furthermore, there is an argument to be made that missile-based weapons are less effective and more expensive than a well-equipped air force (Fetter 1991). Therefore, many nuclear-capable states have foregone proliferation (Paul 2000).

Costs are also an issue. Between 1940 and 1996, the nearly $5,500 billion America spent on nuclear arms and related programmes surpassed government expenditure on health, education, the environment, space research, and law enforcement *combined* (Bidwai and Vanaik 2000: 154). Even while users' return on investment is unclear, some states—even poor ones—persist in their desire to obtain nuclear weapons. With these overwhelming obstacles in mind, Hymans (2006: 8) has stopped wondering why there are so few nuclear states, posing another question instead: "Why are there *any* at all?" (italics original).

Apart from deterrence, Sagan (1996) identifies two other reasons why states develop nuclear weapons. In the "domestic politics model," nuclear weapons become "political tools used to advance parochial domestic and bureaucratic interests" (55). Like the task of playing host to the Games, splitting the atom requires the full mobilization of all the resources a state could possibly call upon. Industry, government, the military, the scientific community, construction and engineering professionals, lay labour, and of course, taxpayers, must be brought on board. With the possible exceptions of waging war, sending people into space, and hosting the Olympics, no other activity of the state requires such a broad coalition of stakeholders. In the Olympic case, we have seen how municipal politicians, property developers, infrastructure conglomerates, and security companies form a powerful stake-holding coalition underwriting the state's Olympic aspirations (Shaw 2008).

Most important to our discussion is Sagan's "norms model" (1996: 73–85). Security constitutes an increasingly large share of what it means to be a modern state. Affirming oneself as a member of the nuclear club is an endeavour to be pursued irrespective of cost–benefit considerations: "Military organizations and their weapons can therefore be envisioned as serving functions similar to those of flags, airlines, and Olympic teams: they are part of what modern states believe they have to possess to be legitimate, modern states" (74). Irrespective of the material resources invested for little or no knowable benefit, O'Neill (2006) reminds us that "states often forgo their direct interests for the sake of prestige, investing in projects that display their modernity, engaging in conflicts over symbols of prestige, or building grand but impractical weapons."

Of course, forbidden fruit taste sweetest of all. Strings of normative regimes have emerged to limit full access to the premier advances of humankind to all but a few nations. The nuclear desire to acquire is foremost among the struggles to level this hierarchy. Even a cursory glance at the speeches of Iranian president Ahmadinejad on the subject of nuclear weapons leaves no doubt that the country's nuclear programme is mobilized through the language of equality and affirmation.

When powerful nations unfairly deny these advancements to those who merit them, they "impose a nuclear apartheid" (Ahmadinejad 2005). The non-proliferation regime only strengthens the hierarchy: a managed system of deterrence on the one hand and a coercively-enforced system of abstinence on the other (Walker 2000) facilitate a stable order that prevents the perils of rapid nuclear armament among great powers (Joseph and Reichart 1998), but prevents lesser powers from joining their ranks.

The Olympics provide a similar mechanism for the formalizing of élite security relationships. Though not nearly as serious in tenor, arguments about who deserves to host the Olympics largely resemble those about who deserves to bear nuclear arms. The success of China's Games went a long way to silencing that nation's critics, showing that China was every bit the modern, advanced state it claims to be, making it worthy of first-class global status. Many feared that Greece would be unable to successfully secure the Olympics, and so Greece initially relied on a coterie of advanced countries to advise them on their security preparations for the Games and to provide substantial material assistance (Brianas 2004: 33–35; Voulgarakis 2005; Government Accountability Office 2005: 3; Lynch and Cuccia 2006: 57). But in spite of these initial concerns, Greek officials were then invited to advise China on their Games in turn. Chinese ascension to Olympus facilitated a quick reconstruction of their international security relationships. Though the US Congress nearly passed a bill condemning the awarding of the Games to China, American officials participated in tactical and intelligence-sharing exercises in support of the Chinese effort.

In such clubs of states, even ostensibly athletic clubs, it should come as no surprise that security demonstrations feature front-and-centre. Architecture, arsenals, and athletes cannot be separated. All three are demonstrative of the broad coalition that only the most competent states can marshal successfully. For its part in affirming claims to membership among the élite club of first-class global citizens, Olympic security cannot be viewed merely as an exaggerated response to credible threats. It must be understood as a parallel manifestation of the excellence and competence that the act of hosting the Games works so hard to demonstrate.

Conclusion

We have argued that material rationalizations for Olympic security provisions exceed the risk rationales through which they are supposedly derived. The Olympics are now hosted in a state of security exceptionalism. We have attempted to denaturalize that claim to security, to show that the differences between normal security provisions and those at the Olympics are born of the stature which host nations garner from the Games. As a security manifestation of the greater Olympic project, deployments of force at the Games demonstrate claims to modernity, technological advancement, wealth, and bureaucratic expertise in the same way that impressive works of engineering or pageantry do. In addition, bold displays of statehood work to carry the positive imagery of Olympic excellence into the political realm. The state's claim to first-class global citizenship is confirmed by its capacity to marshal vast resources to its priorities. This is not to say that the provision of security at the Olympics is wasteful or excessive, but rather that some protected objects are unseen. The Games and the symbolism of prestige are things of great value necessitating a state of exception far greater than that provoked by concerns for human safety or the protection of property, both of which are adequately addressed by the security provisions of normal life.

Still, the Olympics in general remain an enigma. The standards, practices, and judgements of the IOC are largely conducted in secret, and the details of Olympic security, escalated as it is to a state of emergency, remain largely classified. For an event that commemorates universal equality and celebrates the honesty, fraternity, and respect with which we hope that our politics will one day be conducted, the veil of secrecy that envelops the Games is quite tragic. We can only hope that the Olympics, as with all securitizations that have run their course, will at some point return to the realm of the political. We can police them responsibly and celebrate them with the most jubilant vitality, if only the harmonious ethos that was meant for the Games could sneak through the security cordon that surrounds them.

References

Ahmadinejad, Mahmoud. 2005. *Full text of President Ahmadinejad's Speech at General Assembly.* URL: http://www.globalsecurity.org/wmd/library/news/iran/2005/iran-050918-irna02.htm

Barrionuevo, Alexei. 2009. "For Brazil, Olympic bid is about global role." *New York Times*, 27 September.

Beck, Ulrich. 2002. "The terrorist threat: World risk society revisited." *Theory, Culture and Society* 19 (4): 39–55.

Bidwai, Praful, and Achin Vanaik. 2000. *New Nukes: India, Pakistan and Global Nuclear Disarmament*. Oxford: Signal Books.

Boyle, Philip, and Kevin D. Haggerty. 2009. "Spectacular security: Mega-events and the security complex." *International Political Sociology* 3 (3): 257–274.

Brianas, Jason. 2004. NATO, Greece and the 2004 Summer Olympics, Naval Postgraduate School, Monterey, CA. URL: http://handle.dtic.mil/100.2/ADA429691

Buzan, Barry. 2007. *People, States and Fear* (2nd edn). Essex: European Consortium for Political Research Press.

Buzan, Barry, Ole Wæver, and Jaap de Wilde. 1998. *Security: A New Framework for Analysis*. Boulder, CO: Lynne Rienner.

Canadian Air Transport Security Authority. 2007. Summary of the 2008/09–2012/13 Corporate Plan, Capital and Operating Budgets. Ottawa: Government of Canada.

Canadian Security Intelligence Service. 2008. 2010 Vancouver Winter Olympics: Terrorist Threat to Vancouver Area Facilities (unpublished document).

Coates, Dennis, and Brad R. Humphreys. 2003. "Professional sports facilities, franchises and urban economic development," in *University of Maryland Baltimore County, Working Paper Series No. 03-103*.

De Goede, Marieke. 2008. "Beyond risk: Premeditation and the post-9/11 security imagination." *Security Dialogue* 39 (2–3): 155–176.
Degen, Monica. 2004. "Barcelona's Games: the Olympics, urban design, and global tourism." In *Tourism Mobilities: Places to Play, Places in Play*, edited by M. Sheller and J. Urry. London: Routledge.
Duguay, Yves. 2009. "A personal view." *Aviation Security International*, August, 48.
Enders, Walter, and Todd Sandler. 2005. "After 9/11: Is it all different now?" *Journal of Conflict Resolution* 49 (2): 259–277.
Fetter, Steve. 1991. "Ballistic missiles and weapons of mass destruction: What is the threat? What should be done?" *International Security* 16 (1): 5–42.
Goldstein, Lyle J. 2003. "Do nascent WMD arsenals deter? The Sino-Soviet crisis of 1969." *Political Science Quarterly* 118 (1): 53–79.
Government Accountability Office. 2005. Olympic security: U.S support to Athens Games provides lessons for future Olympics. In *Report to Congressional Requesters*. Washington, D.C.
Green, Mary. 2009. *Security Training Unfolds in Vancouver*. [cited 9 February 2009] URL: http://personal.lse.ac.uk/martinak/Olympics/feb23_2009.pdf (accessed December 12, 2010).
Head, M. 2000. "Olympics security: Sydney Olympics used as 'catalyst' for permanent military powers over civilian unrest." *Alternative Law Journal* 25 (4): 192–195.
Hymans, Jacques E. C. 2006. *The Psychology of Nuclear Proliferation: Identity, Emotions and Foreign Policy*. Cambridge: Cambridge University Press.
International Olympic Committee. 2004. *2012 Candidature Procedure and Questionnaire*. URL: http://www.olympic.org/common/asp/download_report.asp?file=en_report_810.pdf&id=810 (accessed March 18, 2009).
Johnson, Chris W. 2006. A brief overview of technical and organisational security at Olympic events. URL: http://www.dcs.gla.ac.uk/~johnson/papers/CW_Johnson_Olympics.pdf
—— 2005. "Using evacuation simulations to ensure the safety and security of the 2012 Olympic venues." *Safety Science* 46 (2): 302–322.
Jones, Harvey. 2009. Telephone conversation with Daniel Bernhard, September 15.
Joseph, Robert G, and John F. Reichart. 1998. "The case for nuclear deterrence today." *Orbis* 42 (1): 7–19.
Kitchen, Peter. 2009. *What the admiral said: Canadian Defence Review interview with vice admiral Dean McFadden* [cited 28 September 2009]. URL: http://www.navy.forces.gc.ca/cms/10/10-a_eng.asp?category=57&id=753 (accessed September 28, 2009).
Klauser, Francisco. 2009. "Spatial articulations of surveillance at the FIFA World Cup 2006™ in Germany." In *Technologies of Insecurity: The Surveillance of Everyday Life*, edited by K. F. Aas, H. O. Gundhus and H. M. Lomell. Abingdon, Oxon: Routledge-Cavendish.
Lynch, Rick, and Phillip Cuccia. 2006. "NATO: Rewarding service in the alliance." *Military Review*, January–February: 54–58.
National Highway Traffic Safety Administration. 2008. *Fatality Analysis Reporting System Encyclopedia* [cited 28 September 2009]. URL: http://www-fars.nhtsa.dot.gov/Main/index.aspx (accessed September 28, 2009).
No 2010. 2009. *Why We Resist 2010* [cited 28 September 2009]. URL: http://www.no2010.com/node/18 (accessed September 28, 2009).

Nye, Joseph S. 1986. *Nuclear Ethics* New York: The Free Press.
O'Neill, Barry. 2006. "Nuclear Weapons and National Prestige." In *Cowles Foundation Discussion Paper No. 1560*.
Owen, Jeffrey G. 2005. "Estimating the cost and benefit of hosting Olympic Games: What can Beijing expect from its 2008 Games?" *The Industrial Geographer* 3 (1): 1–18.
Paul, T.V. 2000. *Power vs. Prudence: Why Nations Forgo Nuclear Weapons*. Montreal and Kingston: McGill-Queen's University Press.
Preuss, Holger. 2004. *The Economics of Staging the Olympics: A Comparison of the Games, 1972–2008*. Cheltenham: Edward Elgar.
Price, Monroe E. 2008. "On seizing the Olympic platform." In *Owning the Olympics: Narratives of the New China*, edited by M. E. Price and D. Dayan, pp. 86–114. Ann Arbor, Michigan: University of Michigan Press.
Sagan, Scott D. 1996. "Why do States build nuclear weapons?: Three models in search of a bomb." *International Security* 21 (3): 54–86.
Samatas, Minas. 2007. "Security and surveillance in the Athens 2004 Olympics: Some lessons from a trouble story." *International Criminal Justice Review* 17 (3): 220–238.
Shaw, Christopher A. 2008. *Five Ring Circus: Myths and Realities of the Olympic Games*. Gabriola Island, British Columbia: New Society Publishers.
Short, J. R., C. Breitbach, S. Buckman, and J. Essex. 2000. "From world cities to gateway cities: Extending the boundaries of globalization theory." *City* 4 (3): 317–337.
Shoval, N. 2002. "A new phase in the competition for Olympic gold: The London and New York bids for 2012 Games." *Journal of Urban Affairs* 24 (5): 583–599.
Statistics Canada. 2008. Motor vehicle accident deaths, 1979 to 2004. In *Health Reports Catalogue no. 82-003-XPE*. Ottawa: Government of Canada.
Thomas, Major Dan. 2010. *Inside Olympic security* [cited 19 March 2010]. URL: http://www.forces.gc.ca/site/Commun/ml-fe/article-eng.asp?id=5829 (accessed March 19, 2010).
Thompson, Drew. 2008. "Olympic security collaboration." *China Security* 4 (2): 46–58.
Tulloch, John. 2000. "Terrorism, 'Killing Events,' and their audience: Fear of crime at the 2000 Olympics." In *The Olympics at the Millennium*, edited by K. Schaffer and S. Smith. London: Rutgers University Press.
Ullman, Richard. 1983. "Redefining security." *International Security* 8 (1): 129–153.
Voulgarakis, George V. 2005. "Securing the Olympic Games: A model of International cooperation to confront new threats." *Mediterranean Quarterly* 16 (4): 1–7.
Wæver, Ole. 1995. "Securitization and desecuritization." In *On Security*, edited by R. Lipschutz. New York, NY: Columbia University.
Walker, William. 2000. "Nuclear order and disorder." *International Affairs* 76 (4): 703–724.
Weldes, Jutta. 1996. "Constructing national interests." *European Journal of International Relations* 2 (3): 275–318.
Weldes, Jutta, Mark Laffrey, Hugh Gusterson, and Raymond Duvall. 1999. "Introduction: constructing insecurity." In *Cultures of Insecurity: States, Communities, and the Production of Danger*, edited by J. Weldes, M. Laffrey, H. Gusterson and R. Duvall. Minneapolis, MN: University of Minnesota Press.
Xiaobo, Liu. 2008. "Authoritarianism in the light of the Olympic flame." In *China's Great Leap: The Beijing Games and Olympian Human Rights Challenges*, edited by M. Worden. London: Turnaround Publishers.

Xu, Xin. 2006. "Modernizing China in the Olympic spotlight: China's national identity and the 2008 Beijing Olympiad." *Sociological Review* 54 (2): 90–107.

Yu, Ying, Francisco Klauser, and Gerald Chan. 2009. "Governing security at the 2008 Beijing Olympics." *International Journal of the History of Sport* 26 (3): 390–405.

Chapter 2

Olympic rings of steel
Constructing security for 2012 and beyond

Pete Fussey and Jon Coaffee

Introduction

The relationship between security and mega-events has received much attention in recent years, particularly after 9/11 where fears of no-warning mass casualty attacks perpetrated by international terrorist actors have stimulated ever more detailed and pre-emptive security responses. These approaches to event security are now becoming relatively standardised as a 'model' or 'blueprint' for reducing vulnerabilities and maximising security at major conferences, cultural festivals and sporting events (*inter alia* Coaffee and Rogers, 2008). Within this context, this chapter examines the form, function and impact of London's 2012 security strategy. It identifies and critiques the role of surveillance as one of its central features and examines the security operation's relationship with prevailing trends evident in previous Olympic and other mega-sporting event security practices.

Despite their plural and locally grounded nature, Olympic-related threats are often exogenously defined (cf. Said, 1993) and, in turn, inspire strong continuities and commonalities across Olympic security responses over both time and place. This paper argues that wider shifts towards 'total' security models comprising continually reproduced security motifs can be observed. These have occurred generally since the terrorist atrocities at Munich (1972) and, particularly, since the International Olympic Committee's (IOC's) more active role in security planning since the 1984 Winter Games in Sarajevo. In turn, these strategies institute Olympic 'spaces of exception' that have become standardised, mobile and globalised. These rebordered spaces ultimately become disassociated from the specific geographical contexts of host cities. The first part of the chapter explores this dynamic in more detail. This is intended to serve three functions. First, it establishes the baseline trajectory of standardised Olympic security strategies onto which London's approach is then mapped. Secondly, contained within this discussion is the argument that these globalised security practices impact unevenly on the idiosyncratic geographies of different Olympic cities to which they are applied. This discord is perhaps most clearly articulated around the theme of terrorism (the central concern of London 2012 security planning). Here, strong commonalities across Olympic security operations contrast with vernacular locally-shaped threats. Finally, given the resonance of previous Olympic security orthodoxies, alongside the formal mechanisms of transferring learning between host cities (see Klauser, this volume), this discussion will locate London's Olympic planning within this much wider, and under-acknowledged, security cycle. In doing so, it is hoped that the scope of London's Olympic security strategy – across bid, preparation, development, application and legacy – is captured.

We argue that London's hosting of the 2012 Olympics both connects with *and* foments different and novel elements to this continuing process of Olympic securitisation. In many respects, the English capital *already* exhibits many of the characteristics comprising standardised Olympic security programmes. Security for 2012 begins at a different point than for other hosts. For example, in addition to the often-cited (and probably outdated) epithet of the world's most surveilled metropolis, London has considerable experience in creating technologically patrolled splintered spaces as a foil to terrorism. After establishing this context, the chapter explores the more specific anatomy of the 2012 project with particular attention to the role of surveillance. This discussion also considers the impact and legacy of the operation. Particularly important here is the contested nature of risk and the legacy of post-event retention of security and surveillance structures. Against a historical backdrop of such measures repeatedly being pioneered on East Londoners, these mechanisms may invest new meaning to the 'community focused' discourse surrounding the 2012 Games.

Olympic (in)securities

Although Olympic-related threats are myriad and complex, after Munich and, particularly since 9/11, Olympic security planning has been dominated by the threat of terrorism. Taking a broad view, terrorist activity around the Olympics has involved myriad forms. Over the last 22 years, for example, these have included perceived threats from left-wing groups (Barcelona, 1992 and Athens, 2004), left-wing state proxies (Seoul, 1988), right-wing extremists (Atlanta, 1996), ethno-nationalist separatists (Calgary, 1988 and Barcelona, 1992), single-issue groups (Albertville, 1992 and Lillehammer, 1994), hostile states (Seoul, 1988) as well as violent Jihadi extremists (Sydney, 2000). Rather than providing an exhaustive and narrative list of events, what follows is a brief analysis of some of the broader processes at work and their implications for future security planning as they apply to selected Olympiads since the late 1980s.

Notwithstanding the prominent and much-discussed attacks at Munich and Atlanta, and despite (and possibly because of) the 'lockdown' of Olympic sites (Coaffee and Fussey, 2010), since 1976 much terrorist activity surrounding the Olympics has occurred outside of the time and place of the event, most notably at the Seoul and Barcelona Olympiads. More recently the 2004 Games in Athens also experienced significant localised terrorist activity in the build-up to their Olympiad – the first post-9/11 Games.

In the case of Seoul, 1988, both the specific geopolitical setting (of the two hostile nations divided by strained border arrangements) and the geographical features of the host nation (that South Korea's *de facto* island status meant visitors overwhelmingly relied on commercial aviation) combined to generate specific terrorist threats against the Games. These threats became manifest when North Korean agents and Japanese Red Army proxies targeted the international aviation industry (United Nations Security Council, 1988). This culminated in the successful bombing of Korean Airlines flight KAL 858 at the cost of 115 lives and a disrupted global campaign against Seoul-bound airlines. Although rooted in the geographical and political contexts of contemporary South Korea, this campaign drew a global response, both in the shape of an unprecedented international Olympic-related intelligence operation and the IOC adopting an enhanced diplomatic function. These have become staples of subsequent Olympic security projects and enabled the IOC to aggrandise its security function and attendant 'knowledge brokering' role (Ericson, 1994).

Spanish authorities faced terrorist threats to the 1992 Barcelona Games from three fronts between 1986 (following the IOC's award of the Games) and 1992. These originated from the left-wing Grupo de Resistencia Antifascista Primo Octobre (GRAPO), Terra Lliure (Catalan separatists), and Euzkadi Ta Askatasuna (ETA). In the run-up to the Games, during 1992 GRAPO conducted five bombings, one small-arms attack and, demonstrating the symbolic value of Olympic targets, conducted a double bombing of a Catalan oil pipeline the day before the opening ceremony (sourced from the Global Terrorism Database (GTD)). More localized still were the actions of Olympic-opposed Catalan nationalists Terra Lliure. Escalated activity comprised 53 attacks (almost exclusively bombings) (sourced from the GTD) including targeting banks sponsoring the Games. The most prominent Spanish group, ETA, simultaneously undertook a sustained bombing campaign culminating in a (media-suppressed) attack on the electricity supply to the opening ceremony (see Toohey and Veal, 2007). ETA's targeting of the Olympics also extended to Madrid's unsuccessful bid to host the 2012 Games. Most directly, this campaign culminated in bombing Madrid's intended Olympic stadium 11 days before the IOC's final decision – or, an attack on the 'symbol of Madrid's candidature' as the Spanish media reported it (*El Mundo*, 2005). Given that different terrorist ideologies influence a varied selection of targets (*inter alia* Drake, 1998; Fussey, 2010 in press), the Olympics provide a ready and consistent symbolic target for myriad groups regardless of their ideological, operational and tactical diversity.

In the 9/11 era of supposedly 'international terrorism' this more localised pattern continues. For Athens, 2004, although, Hinds and Vlachou (2007) argue that, in the main, the Athenian security project was geared towards external threats, domestic threats were most visible. These included the fallout from the conviction of 15 N17 activists during 2003 and the emergence of the anti-Olympic domestic radical group 'Revolutionary Struggle' (who executed five bombings in Athens in the run-up to the Games) (sourced from the GTD).

Olympic threats reconsidered

The above juxtaposition of globalised terrorist risks and the local manifestation of threat articulates a wider theme of Olympic insecurity. Despite the conspicuous internationalism of the Olympic Games, many of the groups that target them are grounded in specific local socio-political contexts (see Fussey *et al.*, 2011 forthcoming, for comprehensive discussion of Olympic threats 1972–2010). Borrowing from Said (1993), Olympic threats should perhaps be considered contrapuntally. The (potentially imperialistic) imposition of an international event harbouring specific values and visions of order stimulates myriad discourses of resistance. Such discourses are generated via complex politico-cultural processes and do not easily lend themselves to be simply cleaved into exogenous and internationalised categories of 'threat'. As Said (1993: 19) notes, 'the world is too small and interdependent' for confrontation to be polarised thus, hence account needs to be made of the confluence of global and local elements, as they constitute Olympic threats. It is in this respect that the 1972 attack at Munich may be seen as truly exceptional, both in terms of its complexity and the importation of activism to the host nation. The state-led murder of 260 protesters 10 days before the 1968 Mexico City Olympics further underlines the novelty of post-1972 emphases on externalised threats.

Olympic security: exceptionality and standardisation

One of the key drivers behind Olympic security programmes, then, has been a 'protectionist reflex' (Beck, 1998: 153) in response to aforementioned generalised and externalised risks. In turn, these risks can be seen to further inform and shape more generalised paradigms of security that have been progressively standardised since Munich. This process has been further underpinned by a shift in the governance of Olympic security throughout the 1980s and the fact that the exceptional nature of such projects entices recourse to previous precedents for which the IOC, as an 'institutional memory', is uniquely positioned to broker appropriate knowledge. This is not to denote a static process, however, as variations of intensity and form (largely depending on the vernaculars of the host's security infrastructures) do exist as does the growing centrality of surveillance within these strategies. Nevertheless, a recurrence and reinforcement of key security motifs can be observed.

Such approaches culminate in Olympic 'spaces of exception' that seek to delineate the Games from their contextual geographies. Comprising these enclaves are myriad security strategies, which, although acknowledged, their exegesis lies beyond the scope of this paper. As such, the role of surveillance is seen as a particularly central feature of these approaches and constitutes the principal site of analysis. To illustrate this process, key components of previous security operations are first outlined to serve as a baseline to examine the relationship to Olympic security processes at subsequent events, including the forthcoming 2012 Games in London.

Responding to Munich: Montreal, 1984 and Seoul policing

Sharply contrasting Munich's 'low-key' approach (reflecting contemporary German sensitivities over conspicuous public displays of social control), little expense was spared on securing the 1976 Olympiad in Montreal. Here, protection from terrorism became *the* key security concern for Montreal's organising committee (COJO). Despite official articulations of 'discreet efficiency' Montreal's strategy was unique in scale and placed enormous emphasis on specific strands of security. Central to these were preventative measures, a strong and visible presence of security forces, and particular emphasis on enhanced and integrated surveillance, communication and decision-making measures (which had been a major failing at Munich and, later, at Atlanta). Specific measures included isolated transport security corridors, enhanced accreditation requirements for site workers and, crucially, probably the first widespread and systematic deployment of CCTV to feature at an Olympics (COJO, 1976). Electronic surveillance was duly given a central role, as articulated in the Official report of the Games:

> it was agreed that the best way to deter suspected trouble-makers was ... [one] that would leave no doubt in their minds they were under continual close surveillance.
>
> (COJO, 1976: 559)

Costing US$100m (equivalent to around US$380m today) such techniques became staple features of subsequent Olympic security strategies and mark the increasing prominence of electronic surveillance.

The baton of technological innovation was transferred to the next Olympiad, the XIII Winter Games at Lake Placid (1980). Here, the most advanced technological measures ever used at an Olympics were deployed. Taking advantage of the particular geography that allowed many Winter Games to be physically separated from their surroundings, many of the innovations involved the surveillance and strengthening of perimeters. In doing so, 12ft-high touch-sensitive fencing, voice analysers, 'bio-sensor' dogs, ground radar, night vision and CCTV were installed (LPOOC, 1980). Together, these represented innovations on surveillance and security practices used to secure military sites and airports and became strategies emulated at subsequent Olympics. Such was the securitisation of this environment at Lake Placid, the legacy of the Olympic village was that it was converted into a correctional facility (*Ibid.*).

This cross-pollination of security also transcended ideological barriers. For example, during 1980, a 'Moscow Doppler' can be observed whereby previous security themes – such as the deployment of US-made security apparatus including metal detectors and x-ray scanners (used at previous Games, including at Lake Placid during the same year) – were incorporated whilst newer approaches were refined – such as zero-tolerance-style policing approaches and exclusion orders – that featured at subsequent Games, notably Sydney and Beijing (albeit with variations of scale). The militarisation of Olympic security has also continued, as evinced (particularly) during the Seoul, Barcelona, Athens and Beijing Games. A further component of these broader strategies has been private security, deployed at Tokyo (1964), refined at Lake Placid and established on a grand scale at Los Angeles (1984).

Olympic security following 9/11

Since 2001, Olympic and major sporting event security strategies have reproduced and built upon these themes. Echoing Ball and Webster's (2004) argument that security antecedents have been intensified, rather than replaced by 9/11, what has shifted since 2001 has been the scale, technological innovation and centrality of surveillance strategies to overall Olympic security planning. Rather than constituting a simplistic expression of technophilia, this commitment to distanciated electronic surveillance is also seen by practitioners to harmonise with the IOC's (often abandoned) aim of projecting the Games as an athletic event and not an exercise in security. This trend is particularly apparent in relation to the Athens and Beijing Olympic security programmes.

Despite being the smallest country to host Olympics since 1952, the Greek Olympics set out the most expensive, elaborate and extensive security programme ever deployed at the Games. Indeed, this first post-9/11 summer Olympiad provides an exemplar (and possibly the apotheosis) of the 'total security' paradigm. Quintupling Sydney's security costs, Greece spent $1.5bn on the Athenian security project (*inter alia* Hinds and Vlachou, 2007). Although partly attributable to a limited extant security infrastructure prior to the Games (particularly when compared to London), much of this cost can be connected to post-9/11 perceptions of vulnerability and heavy commitments to technological surveillance (see Samatas, 2007 and this volume Chapter 3 for authoritative accounts of Athens' exorbitant yet flawed security model).

The IOC's decision to award the XXIX Olympiad to Beijing in July 2001 stimulated a monumental programme of Olympic-related security development. Particularly interesting is how it juxtaposes standardised security practices with the specific confluence of global and national processes at that distinct time. In one respect, Beijing's strategy was facilitated by the state's immense power to mobilise security (as experienced by the totalitarian Moscow and reforming Seoul Olympiads). At the same time, 'Dengist' notions of 'socialism with Chinese characters' (Cook, 2007) – or a state-oriented yet liberalised economy receptive to specific currents of globalisation – enabled the infrastructure and machinery of security to be imported whilst requiring that the global media market export specific brand images of the city.

Technological security measures included embedding Radio Frequency Identification (RFID) tags in some event tickets (such as the opening ceremony) that allowed their holders' movements to be monitored. Despite these headline-catching technologies, principal emphasis rested on more prosaic surveillance camera networks. Initiatives included the 'Grand Beijing Safeguard Sphere', the construction and integration of a city-wide surveillance camera system that some sources (Security Products, 2007) claim cost over $6bn. This and related initiatives has led to estimates that Beijing now hosts over 300 000 public surveillance cameras (*inter alia Los Angeles Times*, 2007). China's recent trend towards hosting international mega-events has also driven surveillance camera deployment across other cities, including Shanghai (hosting 'Expo 2010') and Guangzhou (the 2010 Asian Games). These developments have further catalysed a nationwide 'Safe Cities' programme to establish surveillance cameras in 600 cities (*New York Times*, 2007). Overall, such developments have probably allowed China to claim Britain's dubious accolade of the planet's most intensely observed nation.

Reflecting on Olympic security, 1976–2008

Since 1976, in responding to the risk of asymmetrical and potentially catastrophic terrorist attacks, Olympic security strategies have become increasingly standardised over temporal, national and ideological borders. The components of these strategies can be seen to comprise a heavy commitment to preventative measures, situational crime prevention (particularly in relation to environmental and architectural design), zero-tolerance-style policing, private security, enhanced access controls to Olympic sites, and a central commitment to technological surveillance. Together these inaugurate Olympic imprints – what we might term 'spaces of exception' that are delineated from their host agglomerations. As Agamben (2005: 1) noted in his seminal work on exceptionality, while such conditions are legitimated via their seemingly provisional nature, they operate within 'a no-man's land between public law and political fact'. Once established, such states become sedimented as they 'transform . . . exceptional measure into a technique of government' (*Ibid.* 2). Although referring to conditions of governance, the same may be said for places. Indeed, as Boyle and Haggerty (2009) have noted, such rebordering practices now extend to non-event IOC activities as illustrated by the Committee's recent conference in Guatemala City which resulted in its host neighbourhood being cordoned off and its borders patrolled by armed and militarised police.

This standardisation of practice is also due to extend to future events (as evinced in the familiar components to Rio de Janeiro's 2016 security plan, see Rio 2016). These consistencies also apply to London 2012 as the current Mayor of London, Boris Johnson recently made clear to a Parliamentary Culture, Media and Sport Committee hearing:

> broadly speaking, there will be quite substantial security and protection around the main Olympic venues of the kind that you would expect, and you will be seeing more detail about that nearer the time, but it will be not unlike what they did in China.
>
> (Department for Culture, Media and Sport, 2008)

London, however, introduces a slightly different dynamic to this Olympic security process as the city already boasts considerable experience in constructing sites of exorbitant security (*inter alia* Coaffee, 2009). In contrast to previous Games where the Olympics were seen as a spur and justification for introducing and permanently retrofitting surveillance technologies, London is overtly building on its pre-existing expertise in crime prevention and counter-terrorism in securing the 2012 Games.

Laminating security infrastructures: building on London's track record

As Jennings and Lodge (2009) highlight, for Olympic-type events, security arrangements tend to layer over existing infrastructures (or at least those components that fit the standardised security framework). In the course of over a century of experiencing modern urban terrorism, London has a long history of piloting and subsequently bequeathing advanced surveillance strategies, particularly among its eastern, Olympic-focused territories (Fussey 2007). The city therefore has a mature security infrastructure onto which the 2012 programme will be grafted. Indeed, as the former Olympics Minister, Tessa Jowell, has recently articulated, 2012-related security measures are rooted in the UK having 'years of experience in both tackling terrorism and hosting major sporting and cultural events' (*The Observer*, 2009).

Contrasting the experiences of many host cities (except possibly Atlanta) what is notable about the 2012 site is its development in the heart of an existing urban milieu; one that is densely populated, continually stereotyped through the discourse of 'dangerous spaces' and the host to an overwhelming majority of UK counter-terrorism related investigations. Whilst this segregation of Olympic and non-Olympic venues has been achieved at other Games (with the possible exception of Albertville's sprawling terrain in 1992), this extant urban setting generates additional security challenges. Yet it is perhaps in this respect that London has a particular track record in creating urban enclaves that whilst not physically gated, are symbolically and technologically demarcated from their surrounding environments. Two such examples are particularly prominent in the capital.

High-profile 'spaces of exception'

Following the bombing of London's financial heart in both 1992 and 1993 by the Provisional Irish Republican Army (PIRA), officials created a so-called 'Ring of Steel' which fomented a technologically delineated securitised zone predicated on monitoring and restricting access (Coaffee, 2004). Whilst target-hardening measures (such as security bollards and barriers) altered the urban landscape, it was camera surveillance that was viewed by police as being the most important feature. An additional phase of expansion, intensification and 'hardening' subsequently occurred during the mid-1990s with the introduction of Automatic Number

Plate Recognition (ANPR cameras) and a substantial upgrading of existing camera provisions, rendering the 'Square Mile' area the most intensely monitored space in the UK (Coaffee, 2009), and, in likelihood, Europe. Yet, much of this expansion occurred during a time of reduced threat following the PIRA ceasefire prior to the Docklands bombing in February 1996. Here, the principle drivers were related to a desire to provide abstract (such as 'reassurance') as well as concrete indicators of security to abruptly hesitant investors, tenants and underwriters and avoid an exodus that could threaten London's privileged position within global finance markets (Coaffee, 2004).

Similar principles were engaged and reproduced in the development of the capital's second business and finance metropolis at Canary Wharf in London's Docklands (situated geographically, if not socially or culturally, in the Olympic Borough of Tower Hamlets). Following an averted PIRA bombing in 1992, and a realised attack in February1996, commercial tenants lobbied for a security cordon similar to the Ring of Steel initiated four miles to the west. This fortified landscape or 'Iron Collar' (Coaffee, 2004) was designed along similar security principles and for analogous reasons of reassurance and resilience. As highlighted below, such 'rings of security' bounded by ANPR cameras provide a template for Olympic security planning.

Since the new millennium these cameras have subsequently crept across space and function across the city. ANPR now provides automated sentries around the city-centre perimeter to police London's Congestion Charge. Further afield (and demonstrating the perennial 'Janus face' of surveillance – see Lyon, 1994) these cameras encircle the wider circumference of Greater London's Low Emission Zone to force owners of high-polluting vehicles to be (substantially) charged for using the capital's roads. Despite this diffusion, ANPR has noticeably converged eastwards with the Olympic Boroughs of Hackney (Wells, 2007) and Tower Hamlets (Coaffee *et al.*, 2008) hosting particular concentrations. In the latter case, this technology is notably clustered around Canary Wharf's islands of affluence (coincidentally hosting the headquarters of London's Olympic planners) and its protective 'Iron Collar'.

Surveillance 2.0

Another strong commonality between these splintered spaces has been the use of advanced forms of technological surveillance to patrol their borders and interiors in addition to their broader deployment across the capital. Because these asocial and automated varieties have been previously comprehensively discussed generally (*inter alia* Lyon, 2003) and with reference to London (*inter alia* Coaffee, 2009; Fussey, 2007), the most notable aspects are only briefly revisited here to enable two distinct points of analysis. First, we highlight the propensity for these measures to be deployed into what are generally the poorest areas of London, particularly the three 'Olympic Boroughs' of Tower Hamlets, Newham and Hackney. Secondly, these measures constitute surveillance infrastructures that will be co-opted into the wider 2012 security ensemble, particularly in relation to the more populated areas encircling, and external to, the Olympic Park in Newham.

Most notable of these is Newham council's deployment of Face Recognition CCTV (FRCCTV) throughout the 1990s. Widespread conventional surveillance camera coverage was introduced comparatively late into the area and is more prominently associated with one of the earlier attempts to regenerate Stratford (the heart of 2012-related construction) during 1995's Stratford City Challenge scheme. In an area that had been unable to escape the label of 'dangerousness' that has historically afflicted much of East London, this regeneration stimulated a large deployment of surveillance cameras. FRCCTV followed soon after with the introduction of 300 cameras into the area (*The Guardian*, 2002), arguably one of the first public space deployments of the technology. Officially installed to counter crime and terrorism, in actuality the cameras became operational during 1998, the same year as the 'Good Friday Agreement', which effectively led to the demilitarisation of PIRA. Critics have also emphasised the lack of evidence supporting their effectiveness in tackling crime (see *The Guardian*, 2002).

Other forms of second-generation surveillance strategies – particularly those designed to overcome fallible human attention spans by automatically 'identifying' phenomena deemed suspicious – have also been tested on East Londoners prior to their wider diffusion. One notable example includes the *Intelligent Passenger Surveillance* (IPS) programme that automatically overlays live surveillance camera feeds onto 'ideal' images (such as an empty platform after a train has departed) and alerts of any 'suspicious discrepancies' (such as discarded luggage) first piloted at Mile End Underground station (Tower Hamlets) during 2003. Another less-publicised example was the 2006 experimentation with private sector microphone-equipped cameras in Shoreditch (Hackney) to monitor activity surrounding its thriving night-time economy. Here, manufacturers claimed that they had developed algorithms to distinguish between human screams of pleasure and distress and could automatically alert camera operators to the latter (Wells, 2007) – a claim that convinced municipal 'CCTV Managers' to allow it to be deployed into public spaces.

Overall, this discussion has sought to illustrate the trend of deploying novel forms of technological surveillance into East London, particularly in those areas adjacent to the Olympic Park, and high-profile areas of London more generally. These are the contexts and themes of London's security practice onto which the global leitmotifs Olympic security standards must overlay. At the same time, the (current) £600m 2012 security budget will generate further opportunities to intensify and embed these practices. As the following sections highlight, these trends may forge strong harmonies with the plans to protect the 2012 Games yet, simultaneously, may also create dissonance with some of the idiosyncrasies of policing London.

Securing the 2012 London Olympics

The 2012 Olympic and Paralympic Games will require the largest security operation ever conducted in the United Kingdom. The success of the Games will be ultimately depend[a]nt on the provision of a safe and secure environment free from a major incident resulting in loss of life. The challenge is demanding; the global security situation continues to be characterised by instability with international terrorism and organised crime being a key component.

(Metropolitan Police Authority, 2007)

For host cities, policing the Games is seen as an exercise in exceptionality. The above statement, quoted from a Metropolitan Police (the constabulary covering most of London, including the city's Olympic venues) report on 2012 security planning demonstrates that this view is shared in London. Here, the scale and form of the project constitutes an unprecedented peacetime undertaking for the hosts. Of further interest is the primacy given to securing the Games from a cataclysmic terrorist attack. Indeed, the execution of the July 7 bombings a mere 20 hours after the IOC's decision to award the thirtieth summer Olympiad to London has provided a lasting and symbolic connection between the 2012 Games and terrorist violence. Coupled with unprecedented expenditure following the relaunch of the UK Government's counter-terrorist (CONTEST) strategy and its dominance in local security budgets, the threat of terrorism has become *the* prominent feature of 2012 security planning (Coaffee, 2009). Such prioritisation is not only evident from the above planning document, it is unequivocally articulated in the consolidated *London 2012 Olympic and Paralympic Safety and Security Strategy*: 'the greatest threat to the security of the 2012 Olympic and Paralympic Games is international terrorism' (Home Office, 2009). Moreover, it is important to recognise how the primacy attributed to such concerns shapes the total security infrastructures deployed to secure the Olympics.

Building on the aforementioned foundations, private contractors were invited to construct the substantive and detailed aspects of the strategy via a series of tenders from London's Olympic Delivery Authority (ODA) during 2007. The first round of these tenders enabled private sector projects to be collated into the consolidated 2012 security plan developed and finalised during 2009 (Home Office, 2009) and thus makes it possible to identify the direction of resources within the wider strategy. More specifically, establishing contracts as early as 2007–2008 allowed them to be integrated into the projected Olympic construction programme.

A central theme of this overall security plan was the continuation of London's praxis of fragmenting urban spaces. This is perhaps most clearly articulated by the then Metropolitan Police security coordinator for the 2012 Games' proposed framework for 2012 resilience arrangements. Here, Assistant Commissioner Tarique Ghaffur articulated the need for the Games' counter-terrorism plan to operate over a defined security 'footprint' (territory). In doing so, he evoked the City of London's 'Ring of Steel' as an exemplar of such a security regime (Ghaffur, 2007), one that continues the 'secure by design' mantle informing many prior London developments including Heathrow Terminal Five, the Millennium Dome, Wembley and Lords Cricket Ground. The ODA tenders, designed to enable private security contractors to situate specific strategies within these policy frameworks, reinforce this point. Explicitly characterising the Olympic Park as a splintered 'Island Site' (ODA, 2007), and thus semantically confirming its geographical isolation, these tenders reveal ambitions for technological apparatus to police its borders. In doing so, the ODA placed particular emphasis on procuring technological solutions for these 'problem areas' including 'ACS [access control systems] comprising RFID token and biometric[s]', a 'combination of technology and physical searching' and 'CCTV, security lighting systems and intruder detection systems to be [established,] integrated with, and form a part of, the perimeter security' (*Ibid.*).

These measures also complement more traditional forms of intra-city bordering. The Olympic park itself was 'sealed' in July 2007 and nearby public footpaths and waterways closed for public access. The encircling 11-mile blue fence – 'cordon blue', which was put in place for 'health and safety' reasons, has been likened by some to the Belfast peace walls (*The Guardian*, 2007). In 2009 this was replaced by electric fencing. Around the main venues and on the borders of the Olympic Park there has also been talk of setting up advanced screening access points – the so-called 'tunnel of truth' which can check large numbers of people simultaneously for explosives, weapons and biohazards. Those creating this elaborate security ensemble are also not spared from its controlling features as biometric checks via advanced hand-scanners are routinely carried out on the construction workforce within the sealed site (*inter alia The Observer*, 2009).

To manage this burgeoning assemblage, many disparate technological strategies are pulled together in the Metropolitan Police's newly constructed Special Operations Room, a surveillance camera command centre in Lambeth, South London, inaugurated in 2007. This facility, the largest of its kind in the capital, was ostensibly designed to oversee the security of major events such as the Notting Hill Carnival and large sporting events including the London Marathon and, ultimately, the Olympics. Building on notions of 'nodal governance', this development potentially could be seen in terms of 'nodal security'. Drawing on Castells (2000) characterisation of nodes as the points of confluence for overlapping networks, 'nodal governance' has been developed as a conceptual tool to understand how these nodes then exert their influence (*inter alia* Burris *et al.*, 2005). These sites have been characterised as comprising four features: mentalities (mechanisms for thinking about the governed subject/network); technologies (methods for exercising influence); resources (the means to enable its operation); and institutions (to marshal the mentalities, technologies and resources). The Lambeth control room serves as a conduit, arguably comprising these four elements, where surveillance footage from the control centres of 32 other London boroughs is filtered through. As the network expands, new nodes arise via the forthcoming development of two similar facilities, one in Hendon, North-West London, and another in Bow, in the heart of the East End, close to the Olympic Park. Underlining the prominence of terrorism within this definition of 'security' is the integration of the Metropolitan Police's 'Gold Command' (the largely terrorism-focussed policing body with overall control of strategy and resources) into the strategic and operational management of this facility, their presence also enabling an ability to immediately seize control of the centre during any major incidents.

Partly stimulated by fears of a spatially displaced terrorist attack, these technological security measures will also soak through the 'island site's' borders and permeate the broader locality. As such, during Games time, London authorities will be using advanced surveillance to track suspects across the city including London's ever expanding system of ANPR cameras. A RFID ticket trafficking system, which would allow spectators to be tracked from their home (facilitated by the proposed combined entrance and transport ticket), has also been suggested. Towards the security epicentre in Stratford, a securitised traffic-free 'buffer zone' covering the areas to the south of the Olympic park, from West Ham and Plaistow, has already been approved – a development which may conflict with official declarations that the Games will foster community aggrandisement. Here, such controlled zones may conflict with the area's deep-rooted traditions of (extreme *laissez-faire*) street trading. Additionally, restrictions on protest and assembly around Olympic sites enshrined in the *London Olympic and Paralympic Games Act, 2006* (and likely to be policed by the aforementioned technologies) may potentially stifle another East End ritual: rebellion.

In sum, the confluence of recent security trends in the capital alongside the particular aims of 2012 Olympic security programme creates a climate that elevates the prominence of technological surveillance strategies. Indeed, the ODA's (2007) call for suppliers to offer strategies that 'create an integrated security environment that is effective, discrete and proportionate' echoes the IOC's long-standing aim of projecting the Games as an athletic event and not an exercise in security. What is significant in the context of Olympic security is that 'discretion' and 'proportionality' have often translated into distanciated forms of technological control. This policy direction was further confirmed by the then head of the 2012 security strategy, Tarique Ghaffur, as follows:

> One of the main issues will be technology vs. people ... An event of this scale means technology plays a bigger part in the look and feel of the games and means surveillance will be a major issue that will likely cause debate.
> (Games Monitor, 2007)

Cumulatively, these approaches both harmonise with the standard approaches to Olympic security outlined above and, also, reassert London's tradition of applying technological strategies to 'separated spaces' to tackle crime, terrorism and other risks.

Fanning the flame: security legacies

One of the most frequently articulated concepts in relation to 2012 and a key determinant in London's award of the Games is the issue of 'legacy', particularly in terms of the post-event use of the Olympic site and attendant community regeneration schemes. This post-Games utility also extends to the machinery of security. Indeed, the aforementioned tenders for Olympic Park security providers are encouraging companies to supply 'security legacy', thus bequeathing substantial mechanisms and technologies of control to the post-event site. Here, questions remain over the security priorities of a high-profile international sporting event attended by millions of people and the degree of infrastructure that will remain to police a large urban parkland (the future incarnation of the Olympic site). Although for other mega-events, such as the 2006 FIFA World Cup, the security legacy consisted of sustained networks of professionals rather than physical control measures (largely due to the deployment of mobile surveillance cameras) (Baasch, 2008), this post-event inheritance of security infrastructures is a common Olympic legacy. Indeed, the legacy of retained private policing following the Tokyo (1964) and Seoul (1988) Olympiads and the continuation of zero-tolerance-style exclusion laws after the Sydney (2000) games are a case in point. Private security providers have further expressed the view to the authors that the potential to engage in new markets and develop opportunities with the public sector after the Games is an additional, and perhaps more compelling, reason to engage in the Olympics project. Key here are the critical themes of legitimacy and control 'creep'. Furthermore, issues of citizenship and community (continually cited by the ODA as the main benefactors of the games) may also be called into question.

The social configuration of post-Olympic spaces and their attendant demands for security are also important. A repeated corollary of large-scale redevelopment/regeneration projects generally (*inter alia* Sassen, 2001) and in London specifically (Hall, 2002; Imrie *et al.*, 2009) has been to reinforce micro-community level socio-economic segregation. In turn, the rebordered Olympic neighbourhood is likely to bring a host of new security demands, particularly for surveillance cameras, from the new inhabitants of its gentrified and splintered enclaves, as has traditionally been the case (*inter alia* Sennet, 1996; Bauman, 2000).

Conclusion – Olympic rings of steel

This chapter has argued that despite pluralities of threat and the diverse local topographies that shape them, strong commonalities can be observed across Olympic security operations over both time and place. These catalyse the institution of 'total' security 'spaces of exception' that are simultaneously standardised, transferable and mobile and potentially dislocated from their host environments. Although some localised security vernaculars inevitably penetrate their deployment (particularly given the importance of occupational cultures in delivering security, see *inter alia* Reiner, 2000; Klauser, this volume), the overarching homogeneity of Olympic security arrangements necessarily impact unevenly on diverse host settings. Such asymmetries may relate to issues of efficacy, liberality and applicability.

For London 2012 two additional processes are at play. First, future candidates have noted the currency given by the IOC to London's emphasis on regenerative 'legacies' (see Rio 2016, 2007). Consequently, subsequent Olympics Parks are unlikely to become the suburban appendages of the past and, instead, become hosted within existing urban settings ripe for regeneration. The isolation and bordering of such spaces is likely to become more pronounced. At the same time, the flow of affluent migrants to the wider redeveloped areas may exacerbate demands to retain Olympic security infrastructures. Secondly, prior its Olympic bid, London already boasted a mature security infrastructure comprising key elements of standardised Olympic security. These include the symbolic and technological delineation of urban spaces and piloting advanced surveillance technologies. The £600m (and rising) London 2012 security programme is therefore commencing from a different position than for many preceding Olympic cities and, as the current planning and tendering arrangements demonstrate, is therefore likely to comprehensively intensify and embed the capital's security infrastructure.

This chapter has also argued that Olympic security programmes are largely predicated on externalised terrorist threats. The role of post-millennium tensions informed by partially knowable yet potentially catastrophic risks in shaping 'total security' paradigms also holds for London. Thus, Olympic risks are selectively and socially constructed. The extraction of this notion of threat from the wider canon of (largely human-constructed) contemporary risks (see Beck, 1998) raises a number of final areas of reflection.

Initially, there is potential that standardised security responses may fail to map onto the uneven local topographies of terrorist threats to London. To take the example of one prominent Olympic security component, surveillance cameras, in London they have been more potent in tackling right-wing terrorism than, say, violent Jihadi extremism (Fussey, 2007). An additional consideration is the continual mutability *within* types of terrorist threat. Al Qaeda-inspired activity, for example, has continually shifted from its pre-9/11 (more networked) form and 'post-Taliban' arrangements and is likely to be different in 2012 from what it is at the time of writing.

Accounting for broader social harms generated by the hosting of mega-events further complicates this dynamic. A key issue here is how these (perhaps more routine) harms risk being downplayed by the emphasis on terrorism. Empirical work has demonstrated that hosting mega-events can routinely generate low- and mid-level offences (Decker *et al.*, 2007) as well as organised criminal activity, particularly those of high exploitation (Fussey and Rawlinson, 2009). Indeed, Fussey and Rawlinson's (2009) ongoing ethnographic research into organised crime in East London ahead of the Games, has yielded data revealing the increased mobility (and reduced detectability) of the area's sex industry; transferability of skills as established organised criminal elements enter and consolidate positions in this market; and the establishment of new trans-ethnic coalitions as nodal points to facilitate the industry. Other data from this research reveals the exploitation and theft from vulnerable migrant workers operating below sub-contractor level on construction projects in East London (as part of the wider Olympic-related regeneration programme, although not connected to the actual Olympic Park). In turn, this has generated pathways into the second economy as a strategy of economic survival alongside recourse to violent groups to address fiscal disputes.

As such, numerous criminogenic dynamics may be observed in relation to hosting the Olympics. Here, long-standing recognition that global processes impact on local criminal practices – such as the creation of new entrepreneurially-oriented territories of criminality (*inter alia* Hobbs, 2001) – are germane. Hosting mega-events such as the Olympics accelerate these global processes – including the reweighting of local economies towards (legal and illegal) consumer, leisure and service-oriented markets and their myriad environmental, social and cultural impacts – that, in turn, agitate and impel the development of new criminogenic contexts. Moreover, borrowing from Bauman (1998), such partially visible movements can be seen to impact hidden, vulnerable and transitory populations most acutely and, for those it relocates, mobilise their delineation into protected tourists or policed vagabonds.

At present surveillance-related security dominates security planning for London 2012 given concerns about the threat of international terrorism. Yet the current and long-term impact of non-terrorist criminality, already a significant feature of the Olympic host communities, is likely to grow significantly. Come 2012, the Olympic security operation put in place to protect the Olympic family during the Games – so-called customer-sensitive security – will be intricately blended with strategies which seek to regenerate local communities by upgrading housing, public places and community infrastructures and legacy security. Whereas some elements of the Olympic 'rings of steel' may dissipate in the post-Games era, what remains of the security infrastructure is likely to have significant impact upon everyday life in East London.

References

Agamben, G. 2005. *State of Exception*. Chicago: University of Chicago Press.
Baasch, S. 2008. 'FIFA Soccer World Cup 2006 – Event-driven security policies.' Paper presented at *Security and Surveillance at Mega Sport Events: from Beijing 2008 to London 2012* conference, 25 April.
Ball, K. and F. Webster, eds. 2004. *The Intensification of Surveillance: Crime Terrorism and Warfare in the Information Era*. London: Pluto.
Bauman, Z. 1998. *Globalization: The Human Consequences*. Cambridge: Polity.
––––– 2000. *Liquid Modernity*, Cambridge: Polity.
Beck, U. 1998. *World Risk Society*. London: Polity Press.
Boyle, P., and K. Haggerty. 2009. 'Spectacular Security: Mega-Events and the Security Complex.' *International Political Sociology* 3 (2): 257–274.
Burris, S., P. Drahos, and C. Shearing. 2005. 'Nodal Governance.' *Australian Journal of Legal Philosophy* 30 (1): 30–58.
Castells, M. 2000. 'Materials for an exploratory theory of the network society.' *British Journal of Sociology* 51 (1): 5–24.

Coaffee, J. 2004. 'Recasting the "Ring of Steel": Designing out terrorism in the City of London,' in *Cities, War and Terrorism: Towards an urban geopolitics*, edited by S. Graham, pp. 276–96. Oxford: Blackwell.

—— 2009. *Terrorism, Risk and the Global City – towards urban resilience*. Ashgate: Aldershot.

Coaffee, J., and P. Rogers. 2008. 'Rebordering the city for new security challenges: From counter terrorism to community resilience.' *Space and Polity* 12 (2): 101–118.

Coaffee, J., D. Murakami Wood, and P. Rogers. 2008. *The Everyday Resilience of the City: How Cities Respond to Terrorism and Disaster*. Basingstoke: Palgrave-Macmillan.

Coaffee, J., and P. Fussey. 2010. 'Security and the threat of terrorism.' In *Olympic Cities: City Agendas, Planning, and the World's Games, 1896 to 2012* (2nd edn), edited by J. Gold and M. Gold, London: Routledge.

COJO 1976. *Official Report of the XXI Olympiad, Montreal 1976*, COJO: Ottawa.

Cook, I. 2007. 'Beijing, 2008.' in *Olympic Cities: City Agendas, Planning, and the World's Games, 1896 to 2012*, edited by J. Gold and M. Gold, pp. 286–97. London: Routledge.

Davies, S. 1996. *Big Brother: Britain's Web of Surveillance and the New Technological Order*. London: Pan.

Decker, S., S. Varano, and J. Greene. 2007. 'Routine crime in exceptional times: The impact of the 2002 Winter Olympics on citizen demand for police services.' *Journal of Criminal Justice* 35 (1): 89–103.

Department for Culture, Media and Sport. 2008. *London 2012: Lessons from Beijing*, committee hearing minutes available from http://www.publications.parliament.uk/pa/cm200809/cmselect/cmcumeds/25/8100703.htm.

Drake, C. 1998. *Terrorists' Target Selection*. Basingstoke: Palgrave-Macmillan.

El Mundo. 2005. *ETA pone un coche bomba contra el símbolo de la candidatura de Madrid 2012*, 27 June, available from http://www.elmundo.es/elmundo/2005/06/25/espana/1119719063.html (accessed September 28, 2009).

Ericson, R. 1994. 'The Division of Expert Knowledge in Policing and Security.' *British Journal of Sociology* 45 (2): 149–175.

Fussey, P. 2007. 'Observing potentiality in the global city: Surveillance and counter-terrorism in London.' *International Criminal Justice Review* 17 (3): 171–192.

—— 2011. 'An economy of choice? Terrorist decision-making and criminological rational choice theories reconsidered.' *Security Journal* 24 (1): 85–99.

Fussey, P., and P. Rawlinson. 2009. 'Winners and losers: Post communist populations and organised crime in East London's Olympic regeneration game', presented at the 2nd Annual European Organised Crime Conference, Liverpool, UK, 9 March.

Fussey, P., J. Coaffee, G. Armstrong, and D. Hobbs. 2011. *Securing and Sustaining the Olympic City: Reconfiguring London for 2012 and Beyond*. Aldershot: Ashgate.

Games Monitor 2007. 'The Multinational Security Games', available from http://www.gamesmonitor.org.uk/node/348 (accessed 23 September, 2009).

GCIM, Global Commission on International Migration. 2005. *Migration in an Interconnected World: New directions for action*, Geneva: Global Commission on International Migration.

Ghaffur, T. 2007. *Letter to the House of Lords*, 8 June (London: Metropolitan Police Service).

Guardian, The. 2002. 'Robo Cop.' 13 June, 2002, available from http://www.guardian.co.uk/g2/story/0,3604,736312,00.html (accessed 30 June, 2009).

—— 2007. 'Cordon Blue.' Friday 21 September, available from http://www.guardian.co.uk/society/2007/sep/21/communities (accessed 21 September, 2007).
Hall, P. 2002. *Cities of Tomorrow*. Oxford: Blackwell.
Hinds, A., and E. Vlachou. 2007. 'Fortress Olympics: Counting the cost of major event security.' *Jane's Intelligence Review* 19 (5): 20–26.
Hobbs, D. 2001. 'The firm: Organizational logic and criminal culture on a shifting terrain.' *British Journal of Criminology* 41: 549–60.
Home Office. 2009. *London 2012 Olympic and Paralympic Safety and Security Strategy*. London: HMSO.
Imrie, R., L. Lees, and M. Raco, eds. 2009. *Regenerating London: Governance, Sustainability and Community in a Global City*. London: Routledge.
Jennings, W., and M. Lodge. 2009. 'Governing mega-events: Tools of security management for the London 2012 Olympic Games and FIFA 2006 World Cup in Germany.' ESRC Centre for Analysis of Risk and Regulation Discussion Paper, London School of Economics and Political Science No. 55: 1–25.
Jiminez, F. 1992. 'Spain: The terrorist challenge and the government's response.' *Terrorism and Political Violence* 4 (4): 113–114.
Klauser, F. 2011. 'Commonalities and specificities in mega-event securitization: The example of Euro 2008 in Austria and Switzerland.' In C. Bennett and K. Haggerty eds. *Security Games*. London: Routledge.
Lake Placid Olympic Organising Committee. 1980. *Final Report of the XIII Winter Olympic Games Lake Placid, 1980*, LPOOC: New York.
Los Angeles Times. 2007. 'Beijing Olympics visitors to come under widespread surveillance.' August.
Lyon, D. 1994. *The Electronic Eye: The rise of the surveillance society*. Oxford: Polity.
—— 2003. *Surveillance after September 11*. London: Blackwell.
Metropolitan Police Authority. 2007. Metropolitan Police Service Olympic Programme. Update available from http://www.mpa.gov.uk/committees/x-cop/2007/070201/06/ (accessed 1 October, 2009).
New York Times. 2007. 'China finds American allies for security.' 28 December.
Observer, The. 2009. 'Biometric tests for Olympic site workers.' Sunday 11th, available from http://www.guardian.co.uk/uk/2009/oct/11/biometric-tests-for-olympic-site
Olympic Delivery Authority. 2007. ODA Security Industry Day, Call For Security Tenders, available from http://www.london2012.com/documents/oda-industry-days/oda-security-industry-day-presentation.pdf (accessed 3 December, 2008).
Rio 2016. 2007. *Rio de Janeiro Applicant File – Theme 13: Security*. Rio de Janeiro: Brazil.
Said, E. 1993. *Culture and Imperialism*. New York: Vintage.
Sassen, S. 2001. *The Global City: New York, London, Tokyo*. Princeton, NJ: Princeton University Press.
Samatas, M. 2007. 'Security and surveillance in the Athens 2004 Olympics: Some lessons from a troubled story.' *International Criminal Justice Review* 17 (3): 220–238.
Security Products. 2007. *Beijing To Invest More Than $720 Million For 2008 Summer Olympic Games Security*, available from http://secprodonline.com/articles/2007/08/14/beijing-to-invest.aspx (accessed 10 March, 2010).
Sennet, R. 1996. *The Uses of Disorder: Personal Identity and City Life*. London: Faber & Faber.

Toohey, K., and A. Veal. 2007. *The Olympic Games: A Social Science Perspective*. Wallingford: Cabi.

United Nations Security Council. 1988. *Provisional Verbatim Record of the Two Thousand Seven Hundred and Ninety-First Meeting*, New York: UN, transcribed notes available from http://www.undemocracy.com/S-PV.2791.pdf

Wells, A. 2007. 'The July 2005 Bombings.' Paper given at the Local Government Agency 2007 CCTV Conference: The Development of a National Strategy, Innovative Systems, Effectiveness and Standards, Local Government House, London, 4 July.

Chapter 3

Surveilling the 2004 Athens Olympics in the aftermath of 9/11

International pressures and domestic implications

Minas Samatas

Greece is currently facing a dramatic financial crisis and possible bankruptcy, due not only to international economic conditions, political mismanagement and corruption, but also to the outrageous cost of the Athens 2004 Summer Olympic Games. Contrary to the promises of long-term benefits and progress, these Games had far-reaching negative implications for Greece, especially with regard to its sovereignty and economy. As the nation state which gave birth to the classical and modern Olympics, Greece sought to reaffirm its national pride by hosting the Games in the early twenty-first century. However, the panic surrounding 9/11 was used to force her to accept oversight from foreign security agencies and an extremely expensive but flawed surveillance system.

As the first significant global sports event and mega-media spectacle in the post-9/11 world (Andrews 2003), the Athens Olympics naturally raised very high security concerns (Abe 2004: 224–7), which were based on real and perceived threats from international terrorists. This called for global-scale security coordination and produced enormous security costs for the organizing state (Sidel 2004). Having been planned in the shadow of the 9/11 terrorist attacks and facing international pressure from the US government, the International Olympic Committee (IOC), and the wider "security and surveillance industrial complex" (ACLU 2004), the Athens Olympics were used to promote the latest anti-terrorist technology and served as a showcase for "super-panoptic" surveillance capabilities. These did not ensure the security of the Games, but instead promoted the USA's and other Western nations' economic and political interests. Both of the Greek governments that prepared for and hosted the Games had no choice but to bow to this pressure and buy these expensive surveillance systems, which they saw not only as a way to prevent the Games from being sabotaged, but also as devices that could subsequently be used for crime and traffic control (Samatas 2008).

Although the Games proceeded without terrorist incident, the use of an extraordinarily expensive, and often deficient, surveillance system has proved to have a huge negative impact on Greek sovereignty, fiscal integrity, and civil liberties. As was dictated by Greece's foreign security partners, and especially by the US

Ambassador, the system was used by both Science Applications International Corporation (SAIC) and Siemens as an ambitious experiment in generating profits rather than for producing security and efficiency. This extended to instances of bribery and a scandalous program of tapping the telephones of senior Greek officials. Moreover this hugely expensive Olympic dowry has been used to erode the hard won civil liberties of the Greek people (Samatas 2010).

In another work we have already analyzed the seriocomic fiasco of the Athens Olympics C4I panoptic system, which became a technological disaster for both SAIC and the Greek authorities (Samatas 2007). We have also analyzed the controversy over the post-Games use of the Olympic CCTV cameras and the phone-tapping scandal related to this Olympic security system (Samatas 2008, 2010). The aim of this chapter is to analyze and critique how the USA and other western governments and corporations exploited the 9/11 terrorist attacks to impose their own model of Olympic security and surveillance on Greece to safeguard their own political and economic interests at the expense of Greek sovereignty and the Greek economy.

More specifically, this chapter aims to briefly present and analyze: the extensive post-9/11 international pressures placed on Greece by the international Olympic industry, as exposed in the international media; the imposed international security tutelage and explicit US Olympic intervention in Greek Olympic security; the mandated purchase of a super-panoptic security–surveillance system by a SAIC/Siemens consortium, which caused serious scandals; and the political and economic implications of the above pressures and interests on Greek national sovereignty and the resulting budget deficit. In the context of other pertinent studies which depict the fading glow of the Athens 2004 Olympic Games, we draw some critical conclusions about the issue of Olympic Games surveillance from the perspective of the host nation's sovereignty and economic interests.

We support the findings of this chapter with data from primary and secondary documentary sources: the press, webpages and informal interviews and discussions with politicians, academics, police officers, security personnel, volunteers, and spectators. The chapter also includes revelations from the recent investigations in the German courts and the Greek Parliament into the Siemens C4I scandal, and by the rich international bibliography on Olympic security and surveillance.

Post-9/11 security pressures

Greece, the birthplace of the Olympic Games, won the bid to host the Athens 2004 Summer Olympics in 1997. After the September 11, 2001 terrorist attacks and well ahead of the 2004 Olympics, the Greek government came under tremendous international pressure to ensure the safety of the Games and was mandated to adopt new and intensified counter-terrorism and surveillance protocols (Ball and Webster 2003; Lyon 2003; Gandy 2006; Schulhofer 2002; Sidel 2004; Samatas 2007; Tirman 2004). The American and international media consistently presented negative reports about the security for the Athens Games, helping to construct a global mistrust of Greece's competence in this area, which could have led to the Games being cancelled. In this context, American and international security assistance seemed necessary and even welcome.

Concerted press criticism

Americans were expressing concerns about the Greek government's ability to counter terrorism even before 9/11. For example, in the *New York Times* of July 20, 2001, Greece was described as "one of the weakest links in Europe's efforts against terrorism" (Risen 2000). These perceptions were based on a 2000 Report from the State Department on global terrorism (US State Department 2000). In April 2001, the media again criticized the Greek government for not doing enough to prepare for potential terrorist attacks at the Games (Douglas 2001). Although Al Qaeda had made no specific threat against the Olympics, allegations that "Al Qaeda and its terrorist affiliates have long been known to operate in Greece . . .," made by Philip Shenon in the *New York Times* of August 7, 2003 (Shenon 2003), forced the US State Department to admit that there was "no information to substantiate a verifiable Al Qaeda presence in Greece" (Migdalovitz 2004). Nevertheless, the *New York Times* of July 18, 2004 continued to insist that "while no intelligence agency has picked up any specific threat by al-Qaeda or any other group, no one will deny that Athens is a tempting target" (Bonner 2004). The *Washington Post* of May 6, 2004 also declared, as in many previous reports, that "Athens is at a crossroad to a part of the world where a lot of terrorists come from" (Schmidt 2004), and just before the Games Greek governments were denounced for colluding with domestic terrorist groups, based on alleged "affinities between the radical leftist terrorists and the political elite that emerged after the military junta" (Carassava 2002). Even after the July 23, 2004 arrest and subsequent heavy jail sentences for members of the Greek terrorist group "November 17," the American press continued to suggest that Greece turned a blind eye to terrorism on its soil, reporting that "the biggest concern (is) that some Greek anarchist group will set off a small explosive device in a public area removed from the Olympics, and cause a panic that could affect the Games" (Bonner and Carassava 2004, all quoted by Tsoukala 2006: 46–50).

Just prior to the Games, many international media outlets fixated on the theme that Greece's Olympic efforts were ridiculous. This included accentuating the difficulties the Greek government had in completing the Olympic infrastructure on time and exposing any real or perceived security loophole. Representative British newspaper headlines included "Security the Loser in Athens Race" (Peek 2004), and ". . . Can Greece Turn an Embarrassment into an Olympic Triumph?" (Smith 2004). After the Games a CNN journalist (Reilly, August 31, 2004) admitted: "We were sure every street corner would have three or four terrorists, just kind of killing time, looking for somebody to kidnap. Some bozo said, 'The only place worse to hold an Olympics would be Baghdad.' " Although most US newspapers that assessed the terrorist threat did not openly suggest boycotting the

Games, they also did not exclude the possibility of withdrawing some athletes (Robins 2004).

In contrast to the foreign press, the Greek media defended the Greek Olympic efforts and tried to pacify public opinion. However, long after the Games, the then-Public Order Minister Yiorgos Voulgarakis admitted that security concerns had raised the possibility that the Olympics might not have been conducted in Greece (Gilson 2006). To conclude, as Anastasia Tsoukala (2006: 52) has pointed out, "U.S. press discourse on Athens Olympic security was arguably part of a broader U.S. strategy to impose its own vision of security upon Greece, but also was closely associated with specific U.S. economic and political interests—and the interests of many Western security agencies—whether or not these were relevant to the Olympics."

US Olympic Security Intervention

A US interagency task force was responsible for the Athens Olympics, and was comprised of a security force of 100–110 agents, analysts and administrators from the CIA, the FBI, State, and Defense Departments. All in all, approximately 20 US agencies contributed to the Olympic security effort. The State Department alone spent $2,763,000 to assign 150 special Agents to temporary duty in Athens and surrounding areas prior to and during the Games (Migdalovitz 2004).

The US approach to providing security assistance was based on knowledge gained from Greece's participation in the Department of State's Antiterrorism Assistance Program (ATA) since 1986. According to the US Government Accounting Office Olympic Security Report, the Departments of State, Homeland Security, Defense, and Justice provided security training to various sectors of the Greek government. The US Departments of Energy and Justice provided crisis response assistance during the Olympics, and the State Department also provided special security and other assistance to US athletes, spectators, and corporate sponsors. From March 10th to 23rd, 2004, foreign forces, including 400 US special operations forces, joined their Greek counterparts to deal with multiple terrorism scenarios for suicide bombings, chemical and biological attacks, plane hijackings, hostage situations, and so on. With names like "Trojan Horse," "Gordian Knot," "Hercules Shield," and "Olympic Guardian," such exercises were based on joint operational planning in a simulated Olympic environment in Athens and four other Olympic cities (US GAO Report 2005; Migdalovitz 2004).

The then-US Ambassador to Greece, Thomas Mueller, also played a key role in US interagency efforts to support the Athens Olympics, a role which also helped establish his post-diplomatic career. He directly promoted the SAIC/Siemens security project and after his retirement took a job with a London-based SAIC firm. Acting like an SAIC agent, Mr Mueller stated on April 26, 2002: "This is no time for experiments and testing of new systems to find something better (for Olympics security). Our company has a history and this is a guarantee" (Rizospastis, 2004: 19). The Ambassador was pleased with the results of the joint

American–Greek clampdown in Athens, arguing that "the job here is to put as many locks, sirens and alarms on the house called the Olympics so that the burglar goes to some other house" (O'Neill 2008). Obviously his concern was not only for Olympic security but also for US corporate interests. However, the Ambassador's interference in almost every Olympic matter, his frequent visits to the Greek ministries and his repeated public statements provoked strong reactions from the Greek Parliamentary Opposition. The GAO Report of the US government praised the Ambassador's actions (2005).

Part of the US efforts to strengthen security cooperation long before and during the Athens Games included the FBI and CIA's secret flights for "renditions," which amounted to kidnapping Muslims off the streets of Europe. Although there had been no reports of radical Islamist terrorist groups operating in Greece, the police surveillance of Muslims reportedly had been increased in anticipation of the Olympics. Prior to the Games the CIA made 15 secret flights to Greece: 14 are now known to have occurred in 2002 and 1 just before the 2004 Olympics, according to the MEGA TV Channel programme "Fakeloi" (May 19, 2005). During 2002–2004, CIA agents in Greece kidnapped 12 Muslims who had been targeted by key-word interceptions and other intelligence technologies. Most of these individuals have since been freed, but some have subsequently disappeared (Papahelas, November 25, 2005: A22).

Transnational security training

To ensure the security of the Games, and to appease states that had been critical of Greek security lapses, Greece ultimately surrendered sovereignty to a multinational security consortium. The Greek government signed 38 security agreements with 23 countries to guarantee the security of the Games (Tsoukala 2006: 15). In 2000 the Greek government established a seven-nation "Olympics Advisory Security Team" comprised of the United States, the United Kingdom, Germany, Israel, Australia, France, and Spain, to provide Olympic-related intelligence and training (Samatas 2007). Even NATO, the FBI, CIA and the British M16 were actively involved. Members of this consortium had a headquarters in Athens and participated in training Greek Olympics security forces, focusing on the potential for transnational terrorism. For example, Israeli specialists trained officers to identify and neutralize suicide bombers.

The multinational group also provided advice on technical support issues at the operational level. The range of issues covered included intelligence, planning, training and exercises, technology, command-and-control coordination, and venue security. The United Kingdom chaired the group, which met monthly to coordinate advice and information shared with Greece and assign responsibility for providing security training and equipment. Greece also received security advice from governments not in the Advisory Group, notably Russia, which reportedly sent mobile laboratories to help in the event of a nuclear, biological, or chemical attack and put special forces on standby to deal with a possible threat from Chechen rebels (Migdalovitz 2004).

Greece was forced to build this international security alliance and purchase the latest security and surveillance technology developed by the USA and the European Union to secure their support and confidence, irrespective of the cost or the effectiveness of this technology. According to the then Greek Ministry of Public Order (2004) and now Ministry of Citizens Protection, Greece had initially budgeted US$800 million for Olympic security infrastructure and equipment and assigned some 70,000 military and security personnel to the Games (three times as many as in Sydney and Salt Lake City). NATO provided air cover during the Games with AWACS aircrafts. The International Atomic Energy Agency (IAEA) spearheaded efforts to detect and head off any potential dirty bomb attacks, and the USA provided IAEA with $500,000 for radiation detectors to be used at Olympic events. Further high-level security measures were deployed at the Olympic village, the media villages, the technical officials' villages, Olympic Hotels and the Port of Piraeus, where floating hotels were moored. Athletes and spectators from "high risk" countries, presumably including the United States, Britain, Spain, and Israel, had Greek security escorts. The US and Israeli teams also had their own security forces.

Facing warnings that the Games could be a prime target for international terrorists, there was little domestic opposition to NATO and foreign involvement. Hence, Athens was transformed into an "urban panopticon" or a "fortified urban space" (Coaffee 2003: 15), through an enhanced, sophisticated surveillance security system that promised to shield the Games from international terrorists that might arrive by land, sea or air.

The security industrial complex and SAIC–Siemens

The American Society for Industrial Security (ASIS 2006) claimed that as a result of 9/11, the Athens 2004 Olympics became a testing ground for "the biggest security operation in peacetime Europe." Western corporate security interests are involved in globally promoting sophisticated surveillance systems for mega-events, especially the Olympics, something that is motivated in large part by western corporate security interests, as all these "big sisters" which comprise the "emerging surveillance-industrial complex," have deep ties to the American and E.U. security industries (ACLU 2004: Barlett and Steele 2007; Hayes 2009; 2010).

The Olympic security and surveillance system promoted for the Athens 2004 Olympics caused numerous technical and economic problems during the Games. But there were also legacies during the post-Games period that affected communications security and privacy, national security and sovereignty. These included illegal cellphone wiretaps of government officials, corruption, bribery, and new surveillance cameras which threaten civil liberties. According to recent revelations from German courts investigating the Siemens bribery scandal, the C4I system which was subcontracted to Siemens not only did not work and was far too costly, but was also deployed by Siemens as a "dummy project" to secure business profits, overpayments and bribes rather than as a feasible and effective Olympic security system (Telloglou 2009).

Fierce corporate competition and corruption

Two massive consortia competed for the C4I contract for the 2004 Games. The first was TRS (Thales France and Raytheon System, USA). The other was a consortium comprised of US-based Science Applications International Corporation (SAIC), Siemens (Germany/Greece) and General Dynamics (USA). SAIC is a San Diego-based firm with annual revenues of $7.2 billion and more than 43,000 employees in over 150 cities worldwide, and with close ties to US intelligence agencies. After a highly competitive bidding process lasting 14 months, the Greek Ministry of Defence awarded the contract to build the system, including designing the software and purchasing the necessary hardware, to SAIC on May 1, 2003.

The Greek government chose the SAIC/Siemens bid, despite the fact that the special Greek assessment committee had suggested that the TRS proposal was technically and financially superior. SAIC was chosen ostensibly because it had developed and operated the Sydney 2000 and Salt Lake City 2002 security systems and claimed greater experience with Olympics contracts. The Greek government accepted that: "The SAIC Team is the only team that has direct, hands-on experience integrating and delivering a complete Olympic Security (system)." A rival Raytheon official responded that at Salt Lake City, "SAIC merely plugged into an existing infrastructure that included the CIA, the FBI and the US Armed Forces" (Hadoulis 2002).

The Siemens bribery scandal was one outcome of the fiercely competitive bidding for the lucrative C4I contract. As recently revealed in the US and Germany, the multi-million euro C4I surveillance system, designed by the SAIC corporation and its subcontractor, the German electronics giant Siemens, entailed significant payoffs by Siemens. Documents revealed in the German court show that the Greek branch of Siemens paid over 100 million euros in bribes to Greek politicians and senior officials of both Greek ruling political parties, to ensure that they would be awarded the lucrative Olympic C4I security system and other projects. According to the same sources, the technical complexity and problems of the C4I system facilitated its being used as a "dummy project" to justify overpayments and bribes (Telloglou 2009: 192).

SAIC promised to build the Athenian Olympic "super panopticon" based on the C4I system, which was originally designed for military requirements (Warren 2004). C4I stands for "command, control, communications, and integration" (see www.C4I.org), and this system promised to concentrate the information and management of resources into a vast network of computers, and hundreds of interconnected CCTV cameras all over the Athens metropolitan area, running 24-hours a day. The CCTV systems were linked with a surveillance network of mobile terrestrial trunked radios (TETRAs), which received images and sound in real time and were staffed by 22,160 security personnel and coordinated by a central information security station (*Ta NEA*, 2004: 31). The scope of these requirements was enormous. In addition to covering all of the sporting venues, the safety and security systems also had to monitor nearby harbours, as well as traffic flow in Athens and the surrounding areas. All systems were expected to be used for many years after the Games had finished.

Siemens advertised its C4I system as "the largest and most sophisticated system for civil safety and security applications in the world." Its interconnected systems were supposed to facilitate remote decision-making. The system amounted to a "super-panopticon" (Foucault 1977; Simon 2005; Norris and Armstrong 1999: 222–23) given that its specifications prescribed an electronic nexus of cameras, vehicle-tracking devices and blimps, with continuous online linking of common databases and communication, to provide real-time images and updates of available resources to a central command. It was also designed to integrate mass electronic surveillance with "dataveillance," (Clarke 1994) through data links which would be matched against SIS and Europol blacklists. The system gathered images and audio from an electronic web of over 1,000 high-resolution and infrared cameras, 12 patrol boats, 4,000 vehicles, nine helicopters, a sensor-laden blimp and four mobile command centers. Spoken words captured by sensors connected to the cameras were processed through speech-recognition software which transcribed voice into text that was then searched for patterns, as was other electronic communications entering and leaving the area – including e-mail and image files. Dominic Johnson, Autonomy's chief marketing officer said of his company's software: "It listens, reads and watches. Then it synthesizes. Beyond Greek and English the software understands Arabic, Farsi and all major European languages" (CNN.com 2004).

The C4I Olympic surveillance fiasco

The system was contracted for delivery by May 28, but due to construction delays at some Olympic venues, including the main Olympic stadium, it was delivered just weeks before the opening ceremony. As it was being installed it became apparent that the enormous complexity of the software designed to concentrate the signals from all sensors in a single nerve centre, linked with 30 subsystems, was a technical nightmare (Samatas 2007). For SAIC, marrying the systems to the central Command Decision Support System (CDSS), which forms the backbone of C4I, proved impossible. During the Games, the CDSS continued to crash and was operationally useless, and the system did not function most of the time on any given day. This meant that it was unable to support the 800 staff who were expected to use the system at the main command centre (Psaropoulos 2004a). Greek officials became nervous that they were paying an extraordinary cost for a conventional security apparatus. A US Homeland Security official commenting on the situation joked, "they'll probably end up using their cell phones" (Tsiliopoulos 2004). Nevertheless, Public Order Minister Yiorgos Voulgarakis declared that all the security systems were fully deployed and working smoothly (CNN.com 2004). In fact, the C4I system, which was belatedly contracted, consumed by delays, and untested as of a few weeks prior to the August 13–29 Games, failed to work even long after the Games ended (Hadoulis 2005). By the start of the Games the Greek government had already spent over $750 million on security, almost triple the amount the Australians spent at Sydney in 2000 (Tsiliopoulos 2004: 15).

Given that the C4I did not actually work during the Olympics, the Greek authorities had to rely on more traditional military and policing methods (Samatas 2007; Sugden 2008). The Greek armed forces mobilized some 70,000 soldiers to secure five of the Olympic cities. As a vote of no confidence in SAIC's communications system, these troops used an existing armed forces communication system (Psaropoulos 2004b). As we have concluded elsewhere (Samatas 2007), the fact that the 2004 Athens Olympics had no serious security incidents cannot be attributed to a costly super-surveillance system, but was due to the friendly relations between Greece and both Israel and the Arab nations, as well as to the fact that it was not militarily involved in the Iraq War. Significantly, in late June 2004, Palestinian President Yasser Arafat pledged that Palestinians would adhere to an Olympic truce.

Final acceptance of the SAIC system, which included payment in full, was set for October 1, 2005. However, this deadline was not met, and the government is still negotiating with SAIC to decide how much of the 255 million euro system to accept and pay for. A tentative acceptance of the C4I system by the Ministry of Public Order and SAIC was signed on March 29, 2007. After several modifications in the original agreement, the total cost of the system was decreased from 259.032.250 euros to 245.640.871 euros (estimated at 2003 prices), 75 percent of which has been paid to SAIC (Greek Ministry of Public Order, 2007).

So, six years after the Athens 2004 Games and almost two years after the Beijing Olympics, the Greek government has not yet signed the final acceptance of the C4I system. Its final cost will exceed 250 million euros despite the fact that the C4I system was deficient, did not work during the Olympics, and even after long delays will not include several contracted-for subsystems, now deemed superfluous after the Games. Prospects for a definite settlement are gloomy, since there are judicial investigations into the C4I deal as part of the Siemens bribery scandal. The Greek Ministry of Citizens Protection, which replaced the former Ministry of Public Order under the new PASOK government, submitted a complaint on May 15, 2010 against the SAIC company for not installing the C4I security system on time and not training Greek personnel to use it. Given the lack of staff trained in how to handle the system, it has essentially become useless. The ministry is seeking compensation from SAIC for between 80 and 100 million euros. SAIC has also filed a complaint to the International Court, which will decide on this matter (GRReporter, 2010).

Implications for national and economic sovereignty

Greeks are extremely sensitive about their national sovereignty, so requests for international assistance with security for the Olympics were viewed as potentially politically explosive. Nonetheless, Greece, a small and financially weak state, and the first country to host the Summer Olympics after 9/11, was obliged to surrender sovereignty to a multinational security consortium and to accept foreign interference.

To pacify Greek public opinion the Greek government exercised "rhetorical sovereignty" during the Games, by claiming that Olympic security was the responsibility of the Hellenic Police (Ministry of Public Order) and specifically of the Olympic Games Security Division (OGSD), a special police unit created for the 2004 Olympic Games. They also announced that the multinational Olympic Security Advisory Group would report to the Greek Minister of Public Order on security issues at the strategic level. Further, authorities stressed that although foreign secret service agents would protect certain VIPs, Greek security officials prohibited them from carrying weapons within Greece, as it is against Greek law (Tsoukala 2006: 52).

Despite this "rhetorical sovereignty," Greece was under international security tutelage to meet the security requirements of the eight-week-long Olympics and Paralympics by virtue of how they had to underwrite a very expensive and ultimately ineffective surveillance system. Greece had to sacrifice its sovereignty to pacify the international community's post-9/11 terrorist panic (Tirman 2004), and forge a security partnership with the USA and their allies (Samatas 2007). Regarding the choice of SAIC and its costly C4I system, as one foreign diplomat put it: "The Greek government was subliminally obliged to hire a US firm because the US was breathing down the necks of the Greeks on security. By hiring a US firm Greeks could turn around and say, 'You guys are in charge. Don't blame us' " (Psaropoulos 2004b). Greeks have been "asked to pay the sum of our fears" stated the *Washington Post* one day before the opening of the Games, clearly underscoring the fact that the major part of the Olympic security apparatus had resulted from US pressure (Tsoukala 2006: 14). Long after the Games, then-Public Order Minister Yiorgos Voulgarakis responded to opposition criticism that Greece acceded to foreign pressures regarding security for the 2004 Games by observing: "I did not open my mouth . . . because I realized that by speaking out the Olympics might not have been conducted in Greece and our relations with foreign countries would be torpedoed" (Gilson 2006).

The C4I-related phone tapping scandal was a further blow to Greek sovereignty. The Government revealed in 2005 that the phones of Greek Prime Minister Kostas Karamanlis, the ministers of foreign affairs, defence, public order, justice and many other top government, military and security officials were tapped by unknown individuals during the Athens 2004 Olympics and for nearly a year after, until these intercepts were discovered on March 7, 2005. So far it has not been possible to identify who was behind the unlawful surveillance (Prevelakis and Spinellis 2007). Our research has led us to adopt the dominant theory advanced by former US diplomat J.B. Kiesling that the phone taps were organized by the US secret service agencies for reasons related to American Olympic security concerns, the C4I problems, and the US mistrust of the Greek government and its

Olympic security system (Samatas 2010). Kiesling (2006) argued that, "by bugging more Greek Olympic security officials than local radicals, the eavesdroppers fuelled speculation that they were less concerned for the safety of athletes and spectators than for the fortunes of SAIC."

The extraordinary security–surveillance cost

Greece has spent hundreds of millions of dollars on security technology as part of its larger $1.5 billion tab to ensure the safety of athletes, spectators and others. As one diplomat concluded: "Greece, a country of 11 million people is spending 300 million dollars to buy what is the Cadillac of security systems, when what it can really afford is the Chevrolet" (Psaropoulos 2004b: 18). But Greece alone has been left to pay the $1.5 billion bill, an amount nearly triple the original projections. This was more than four times the security costs for the 2002 Winter Games in Salt Lake City and six times that spent for the 2000 Summer Games in Sydney. The government's budget deficit, as a percentage of gross domestic product, grew from 4.6% in 2003 to an estimated 5.3% in 2004. The entire cost of the Athens Olympics has risen to $15 billion. Olympic spending left Greece with a hefty budget deficit of 6.1% of the Gross National Product in 2004, breaching the EU's 3% cap. The higher Olympic costs also pushed up public borrowing in 2004 to 43 billion euros from an earlier target of 35 billion. Since then, public borrowing has exploded and in 2008 was 107% of the GNP (Papadimas 2005; Itano 2008).

In a prescient observation after the Games, Rick Reilly of *CNN Sports Illustrated* wondered:

> Why you had to pay for our paranoia, I'll never know. It is the world's problem; the world should have to pay for it. What small country is going to be able to afford to host the Olympics anymore with these insane security demands? From now on, if a country wants to send a team to the Games, it pays its share of security, based on its share of the gross world product. In other words, it's our war; we should have to pay for it . . . (Reilly 2004).

Concluding remarks: the fading glow of the Athens Olympics

Assessments of the Athens Games were largely complimentary and even apologetic. For instance, the *San Francisco Chronicle* concluded: "A world that feared the worst got the best. The Athens Olympics supplied a remarkable array of feats, drama and spectacle. The end result made it hard to remember the doubts. Terrorism? After $1.5 billion was spent on security, the worst that happened was a spectator in a skirt rushing at a marathon runner on the final day . . . the Games were a sports fan's dream" (*San Francisco Chronicle*, 2004: B8). Sally Jenkins of the *Washington Post* proposed the following: "Let's Give These Games A Gold Medal. In summing up the Athens Games, the first order of business is to extend a big 'sorry' to the Greeks. Nothing blew up and nothing collapsed, and nothing less has been accomplished than the full restoration of Athens as a splendid world capital. The Greeks have proved a very pointed point. There is more than one way to throw an Olympics . . . The Greeks did this in their way—while also humouring the stubborn American fixation on security, our conviction that our personal safety matters most and our definition of danger is the only one that counts . . ." (Jenkins 2004). Rick Reilly of *CNN Sports Illustrated* observed, "We insisted you spend 1.2 billion euros on security. You had to put up blimps and cameras all over the city. You couldn't throw a bucket of grapes anywhere and not hit a soldier with a rifle. And nothing happened. Zero" (Reilly 2004).

Following the 9/11 terrorist attacks, real and perceived security concerns combined with pressure from various political and economic interests in the USA, the Western international community, and the Olympic Industry forced Greece to finance an extraordinary security and surveillance project, far beyond the original plan proposed in 1997. This system was ultimately deficient and financially unbearable. Despite the Greek government's rhetorical façade of sovereignty during the Games, Greece had actually surrendered sovereignty by accepting the security tutelage of a multinational security consortium and accepted US overt interference in order to avert threatened boycotts and ensure the Games' security. That enormous Olympic security and surveillance system produced an expensive Olympic surveillance "dowry" of hundreds of CCTV cameras, which citizens perceive to be a threat to hard-won Greek civil liberties; a serious bribery scandal perpetrated by Siemens to promote its C4I system; a prolonged phone-tapping program that targeted the Greek government, contravened its sovereignty and democracy; and a skyrocketing public debt, a major cause of the Greek national bankruptcy.

The multinational corporations, which produce the high-priced and immensely profitable security and surveillance systems (Hayes 2009), compete fiercely at every mega-event to secure lucrative contracts. Payoffs and bribes to politicians and state officials, and various other unlawful activities have become commonplace strategies to ensure large profits, as evidenced by Siemens bribing officials to win the competition to install the systems for the Athens 2004 Games. But there is also a longer history of such corruption. For example, the Australian Olympic Committee (AOC) president John Coates revealed that he, and other officials, had been involved in extensive vote buying in 1993 to secure Sydney's Games' bid (Phillips 1999). In China, the official in charge of building Beijing's Olympic Games venues, worth about $55 billion, was sentenced to death, for taking $1.45 million in bribes, a penalty which was suspended for two years and then commuted to life imprisonment (Callick 2008).

The illegal cell-phone monitoring, before, during and long after the Athens 2004 Olympic Games, which targeted the Greek prime minister, his government and top military and security officials touched not only on the privacy of communications and human rights, but also on sensitive areas of national security and sovereignty. This further escalated the heavy costs Greece paid for these Games.

The Athens Olympic "surveillance Games" also raise two important issues regarding the efficiency and the economic cost of surveillance systems for mega-events. The fact that the surveillance system proved to be deficient and unworkable, even long after the end of the 2004 Olympics, and that Olympics security was achieved mostly by conventional military and security methods, raises serious scepticism about the real anti-terrorist efficiency of high-tech surveillance systems.

Regarding costs, Greece bankrupted itself to host a successful Olympiad (Samatas 2007). Such spending came at the expense of social programming. Meanwhile the legacies of restricted civil liberties have increased post-Olympic skepticism and anti-surveillance concerns in Greece. With most of the Olympic venues now closed, empty and decaying, Olympic scepticism among Greek citizens is only reinforced. To avoid a future where only wealthy nations host the Olympics we must look for initiatives where the international community would share security costs at future Olympics, especially when they occur in financially weak states.

In sum, the troubled example of the Athens 2004 Olympics allows us to conclude that 9/11 accelerated the transformation of the Olympics into an enormously costly security and surveillance spectacle (Boyle and Haggerty 2009) that benefits powerful international political and economic interests at the expense of the hosting state's sovereignty and economy. While the Athens 2004 Olympics boosted the morale of the country that is the birthplace of the Olympics and fostered a sense of pride amongst Greek citizens that their small poor country could organize and host a successful Olympiad, the 2004 Olympics had an overall detrimental cost for Greece. While much of this can be attributed to the meddlesome influence of foreign economic and security interests, it is also the case that Greece's political and structural inefficiencies hampered the country's ability to translate the Games into lasting momentum towards modernization. Now it must pay the painful bill.

References

Abe, K. 2004. "Everyday policing in Japan: Surveillance, media, government and public opinion." *International Sociology* 19 (2): 215–231.

American Civil Liberties Union (ACLU). 2004. The surveillance industrial complex. URL: www.aclu.org/SafelandFree/ (accessed October 9, 2009).

American Society for Industrial Security (ASIS). 2006. ASIS January 2006 Newsletter San Diego Chapter. URL: www.asissandiego.org/newsletters/pdffiles/2006_01.pdf (accessed October 2, 2009).

Andrews, D. 2003. "Sport and the transnationalizing media corporation." *Journal of Media Economics* 16 (4): 235–252.

Ball, K. and F. Webster, eds. 2003. *The Intensification of Surveillance.* London: Pluto.

Barlett, D. L. and J. B. Steele. 2007. "Washington's $8 billion shadow." *Vanity Fair* March 2007. URL: http://www.vanityfair.com/politics/features/2007/03/spyagency200703 (accessed June 24, 2010).

Bonner, R. 2004. "Providing security, with the help of some friends." *New York Times*. July 18. URL: http://www.nytimes.com/2004/07/18/travel/providing-security-with-the-help-of-some-friends.html (accessed July 30, 2009).

Bonner, R. and A. Carassava. 2004. "Pushed by U.S., Greece to allow troops at Olympics." *New York Times*. July 21. URL: http://www.aliraqi.org/forums/archive/index.php/t-34184.html (accessed December 4, 2010).

Boyle, Philip and Kevin D. Haggerty. 2009. "Spectacular security: Mega-events and the security complex." *International Political Sociology* 3 (3): 257–74.

Callick, R. 2008. "Beijing Olympic Games bribe official Liu Zhihua on death row." *The Australian*. October 21. URL: http://www.theaustralian.com.au/news/bribe-official-on-death-row/story-e6frg6t6-1111117804370 (accessed November 2, 2009).

Carassava, A. 2002. "Arrests destroy noble image of guerrilla group in Greece." *New York Times* July 23. URL: http://www.nytimes.com/2002/07/23/world/arrests-destroy-noble-image-of-guerrilla-group-in-greece.html?scp=1&sq=affinities%20between%20the%20radical%20leftist%20terrorists%20&st=cse (accessed August 13, 2009).

Clarke, R.A. 1994. "The digital persona and its application to data surveillance." *The Information Society* 10 (2): 77–92.

CNN.com. 2004. "Olympics' digital security unprecedented." by Marauder Wednesday, August 11, 9.24 am. URL: http://venus.cs.aueb.gr/forum/viewtopic.php?f=45&t=614&start=0 (accessed December 4, 2010).

Coaffee, J. 2003. *Terrorism, Risk and the City*. Aldershot: Ashgate.

Douglas, F. 2001. "Olympics; More to it than games." *New York Times* July 24. URL: http://www.nytimes.com/2001/07/24/sports/olympics-more-to-it-than-games.html?scp=3&sq=Olympics%3B+More+to+it+than+games&st=nyt (accessed December 4, 2010).

Foucault, M. 1977. *Discipline and Punish: The Birth of the Prison*. New York: Vintage Books.

Gandy, O. Jr. 1989. "The surveillance society: Information technology and bureaucratic social control." *Journal of Communication* 39 (3): 61–76.

Gandy, O. Jr. 2006. "Data Mining, Surveillance, and Discrimination in the Post-9/11 Environment." In *The New Politics of Surveillance and Visibility*, edited by K. Haggerty, and R. Ericson, 363–384. Toronto: University of Toronto Press.

Gilson, G. 2006. "ND denies it used EYP to spy" *Athens News*, March 31. URL: http://www.athensnews.gr/old_issue/13176/14324 (accessed December 4, 2010).

Greek Ministry of Public Order, Hellenic Police. 2004. "Olympic Security Planning," URL: http://www.astynomia.gr/images/stories/Attachment13907_o_asfalia_eng.pdf (accessed May 15, 2010).

Greek Ministry of Public Order. 2007. Press release, March 29. URL: http://www.yptp.gr/index.php?option=ozo_content&lang=GR&perform=view&id=2097&Itemid=360 (accessed September 4, 2009).

GRReporter, 2010. "Expensive and useless for the Greeks turned out to be the C41 security system." May 26. URL: http://www.grreporter.info/en/expensive_and_useless_greeks_turned_out_be_c4i_security_system/2853 (accessed May 27, 2010).

Hadoulis, J. 2002. "Athens security suitors bid to face the unthinkable in 2004." *Athens News*, October 25. URL: http://www.athensnews.gr/old_issue/12985/8850?action=print (accessed December 4, 2010).

Hadoulis, J. 2005. The magic number that does everything" *Athens News*, February 18. URL: http://www.athensnews.gr/old_issue/13118/12640 (accessed December 4, 2010).

Haggerty K. and M. Samatas eds. 2010. *Surveillance and Democracy*. London: Routledge.

Hayes, B. 2009. *NeoConOpticon: The EU Security-Industrial Complex*. TNI, Transnational Institute and Statewatch.

——— 2010. "Full Spectrum Dominance' as European Union security policy: On the trail of the NeoConOpticon." In *Surveillance and Democracy*, edited by K. Haggerty and M. Samatas, pp. 148–69. London: Routledge.

Itano, N. 2008. "As Olympic glow fades, Athens questions $15 billion cost." *The Christian Science Monitor*, July 21. URL: http://www.csmonitor.com/World/2008/0721/p04s01-wogn.html (accessed August 10, 2009).

Jenkins, S. 2004. "Let's Give These Games A Gold Medal." *Washington Post* August 30, p. D01. URL: http://www.washingtonpost.com/wp-dyn/articles/A45135-2004Aug29.html (accessed November 20, 2009).

Kiesling, J. B. 2006. "An Olympian scandal." *The Nation*, March 20. URL: http://www.thenation.com/article/olympian-scandal (accessed June 24, 2010).

Lenskyj, H. J. 2000. *Inside the Olympic Industry*. New York: SUNY Press.

——— 2004. "Making the World Safe for Global Capital: The Sydney 2000 Olympics and beyond." In *Post-Olympism? Questioning Sport in the 21st Century*, edited by J. Bale and M. K. Christensen, pp. 135–45. New York: Berg.

——— 2008. *Olympic Industry Resistance*. New York: SUNY Press.

Lyon, D. 2003. *Surveillance after September 11*. London: Polity.

Migdalovitz, C. 2004. Greece: Threat of terrorism and security at the Olympics. Congressional Research Service (CRS) Report for Congress. URL: http://www.fas.org/irp/crs/RS21833.pdf (accessed September 15, 2009).

Moechel, E. 2008. "The surveillance Olympics, Beijing 2008 – powered by European technology." Brussels: Olympic Rights for Human Games Conference. May 15. URL: http://www.quintessenz.at/doqs/000100004353/2008_05_11_olympic_surveillance.pdf (accessed December 4, 2010).

Norris, C. and G. Armstrong. 1999. *The Maximum Surveillance Society: The Rise of CCTV*. Oxford: Berg.

O'Neill, B. 2008. "Double standards are no friend of freedom." *Spiked*. July 31. URL: http://www.spiked-online.com/index.php/site/article/5532/ (accessed November 20, 2009).

Papadimas, L. 2005. "Athens counting cost of the Olympics." Sport.Sportman.com. August 4. URL: http://sport.scotsman.com/athensolympics/Athens-counting-cost-of-the.2648827.jp (accessed August 10, 2009).

Papahelas, A. 2005. "Secret CIA flights from Athens." TO BHMAonline, November 27. URL: http://www.tovima.gr/default.asp?pid=2&ct=78&artid=169713&dt=27/11/2005 (accessed August 12, 2009).

Peek, L. 2004. "Security the loser in the Athens race." *The Times* (London), April 13.

Phillips, R. 1999. "Sydney revelations deepen Olympics corruption scandal." *World Socialist Web Site* January 30, URL: http://www.wsws.org/articles/1999/jan1999/olym-j30.shtml (accessed November 2, 2009).

Pound, R. W. 2004. *Inside the Olympics. A Behind-the-Scenes Look at the Politics, the Scandals, and the Glory of the Games*. Etobicoke Ont: John Wiley & Sons, Inc.

Prevelakis, V. and D. Spinellis. 2007. "The Athens Affair." *Spectrum*, Institute of Electrical and Electronics Engineers (IEEE), 44 (7): 26–33.

Psaropoulos, J. 2004a. "SAIC and government discuss a change of security plans." *Athens News*, May 7. URL: http://www.athensnews.gr/old_issue/13065/11242 (accessed December 4, 2010).

—— 2004b. "SAIC agreement up in air." *Athens News*, July 23. URL: http://www.athensnews.gr/old_issue/13076/11553 (accessed December 4, 2010).

Real, M. 1996. "The postmodern Olympics: Technology and the commodification of the Olympic movement." *Quest* 48: (1): 9–24.

—— 2004. "The political economy of the Olympic Games: History and critical theory of modern Games" (in Greek), *Zitimata Epikoinonias* (Communication Issues), 1 (1): 7–23.

Reilly, R. 2004. "We were Wrong." *CNN Sports Illustrated*, August 31. URL: http://sportsillustrated.cnn.com/2004/olympics/2004/writers/08/29/reilly.letter/index.html. (accessed June 24, 2010).

Risen, J. 2000. "A Pattern of Unsolved Greek Terrorism Cases." *New York Times*, June 9. URL: http://www.nytimes.com/2000/06/09/world/a-pattern-of-unsolved-greek-terrorism-cases.html (accessed August 3, 2009).

Rizospastis. 2004. SAIC like CIA, FBI NATO. May 14. URL: http://www1.rizospastis.gr/wwwengine/story.do?id=2321940&publDate=14/5/2004 (accessed July 15, 2009).

Robins, L. 2004. "Olympics; security at summer games in Athens is topic A, B and C for U.S. basketball." *New York Times*, April 29. URL: http://www.nytimes.com/2004/04/29/sports/olympics-security-at-summer-games-in-athens-is-topic-a-b-and-c-for-us-basketball.html (accessed August 7, 2009).

Samatas, M. 2004. *Surveillance in Greece: From anticommunist to the consumer surveillance*. New York: Pella.

—— 2007. "Security and Surveillance in the Athens 2004 Olympics: Some lessons from a troubled story." *International Criminal Justice Review* 17 (3): 220–238.

—— 2008. "From thought-control to the traffic-control: CCTV politics of expansion and resistance in post-Olympics Greece." In *Surveillance and Governance: Crime Control and Beyond*, edited by M. Deflem, pp. 345–69. Bingley: Emerald.

—— 2010. "The Olympic phone-tappings in Greece: A defenceless state and a weak democracy." In *Surveillance and Democracy*, edited by K. Haggerty and M. Samatas, pp. 213–30. London: Routledge.

San Francisco Chronicle. 2004. "Grecian Formula." August 31. URL: http://www.sfgate.com/cgi-bin/article.cgi?f=/c/a/2004/08/31/EDG2E8FVCS1.DTL (accessed August 12, 2009).

Schulhofer, St. J. 2002. *The Enemy Within: Intelligence Gathering, Law Enforcement, and Civil Liberties in the Wake of September 11*. Washington D.C.: Twentieth Century Fund.

Schmidt, S. 2004. "New Fears About Olympics; Blasts in Athens Heighten Security Concerns as Games Near." *The Washington Post*, May 6. URL: http://pqasb.pqarchiver.com/washingtonpost/offers.html?url=%2Fwashingtonpost%2Faccess%2F629193951.html%3FFMT%3DFT%26FMTS%3DABS% (accessed December 4, 2010).

Science Application International Corporation (SAIC) Magazine 2003. SAIC wins IT security contract for 2004 Athens Olympics. URL: http://www.saic.com/news/saicmag/2003-summer/olympics.html (accessed July 20, 2006).

—— 2006. "A gold-medal achievement: SAIC and the 2004 summer Olympic Games in Athens." *American Society for Industrial Security*. URL: http://www.asis-sandiego.org/newsletters/pdffiles/2006_01.pdf (accessed October 2, 2009).

Shenon, P. 2003. "Threats and Responses: Terrorism; U.S. Is Inspecting Overseas Airports for Missile Threats." *New York Times*, August 7. URL: http://www.nytimes.com/2003/08/07/world/threats-responses-terrorism-us-inspecting-overseas-airports-for-missile-threats.html (accessed July 30, 2009).

Sidel, M. 2004. *More Secure Less Free? Antiterrorism Policy and Civil Liberties after September 11*. Ann Arbor: University of Michigan Press.

Simon, B. 2005. "The return of panopticism: supervision, subjection and the new surveillance." *Surveillance & Society* 3 (1):1–20.

Smith, H. 2004. "Clock ticks as Athens sprints for the line: With less than four months to go, can Greece turn an embarrassment into an Olympic triumph?" *Guardian*, April 22.

Sugden, J. 2008. "Watching the Games." Washington DC: *Foreign Policy In Focus*. August 22.

Ta NEA. 2004. Report (unnamed), March 20: 31.

Telloglou, T. 2009. *To Diktyo (The Network)*. Athens: Sky.

Tirman, J. ed. 2004. *The Maze of Fear: Security and Migration after 9/11*. New York: The New Press.

Tsiliopoulos, E. 2004. "SAIC struggles to meet deadlines." *Athens News*. March 5. URL: http://www.athensnews.gr/old_issue/13056/10992 (accessed December 4, 2010).

Tsoukala A. 2006. "The security issue at the 2004 Olympics." *European Journal for Sport and Society* 3 (1): 43–54.

United States Government, Accountability Office (GAO). 2005. *Olympic Security: U.S. Support to Athens Games provides lessons to future Olympics*. GAO-05-547, A report to congressional requesters, Washington D.C. URL: http://www.gao.gov/new.items/d05547.pdf (accessed November 23, 2009).

U.S. State Department, Office of the Coordinator for Counterterrorism. 2000. Europe overview: Patterns of global terrorism 2000. April 30. URL: http://www.state.gov/s/ct/rls/crt/2000/2434.htm (accessed August 1, 2009).

Warren, R. 2004. "City streets – the war zones of globalization: Democracy and military operation on urban terrain in the early twenty-first century." In *Cities, War and Terrorism*, edited by S. Graham, 214–230. London: Blackwell.

Chapter 4

The spectacle of fear
Anxious mega-events and contradictions of contemporary Japanese governmentality

David Murakami Wood and Kiyoshi Abe

Introduction

This paper is one of a series of papers compromising an empirically informed re-theorization of surveillance and social order in Japan, which aim to contribute to several different debates. In the first we, together with David Lyon, set out a critical historically grounded introduction to surveillance in the context of Japanese cities (Murakami Wood et al. 2007). In this second paper, through a history of mega-events in post-war Japan, we reconsider the relationship of Japan to the "outside," from where the Japanese state has long drawn its particular conception of "Others" who must be either excluded, monitored or otherwise governed to ensure social order. We stress the anxiety that results from increasing openness and the feeling of state inferiority that comes from the ongoing effort to establish and intensify a certain (and ever-changing) status as an "acceptable" global power. We suggest in this paper, in what we believe is an important correction to conventional interpretations of this relationship, that this effort has significant effects not just on international relations (with which this paper is not concerned), but also on domestic governmentality. In the third paper (Murakami Wood and Abe forthcoming), we continue to use the example of mega-events, but expand on those domestic, and urban, governmental contexts, characteristics and effects, with particular attention to the changing "aesthetic" of social order in Japan.

Why mega-events? Mega-events are an exemplary interface between global and local or national, and form a crucible of governmental anxieties. These anxieties may be temporarily appeased by the security show or stage-set security that accompanies such events, but their more important function for Japan, and perhaps more generally, has been their signaling of the emergence of a new global form of surveillant governmentality, a global mode of ordering, and part of a ready-made governance "package" that goes with the neoliberal global economy. They thus form a highly instructive set of cases from which the relationship between external and internal modes of social ordering can be examined.

We first introduce the argument in a recent historical context, which draws upon Masachi Ohsawa's tripartite periodization of Japan's post-WW2 ideological orientation (Ohsawa 1998; 2008). We then consider the FIFA World Cup 2002 and its lineage from previous sports mega-events, before moving onto how the traveling circus of new global governance, like in this case the G8 summits, express similar trends and problems. This is tied together with a consideration of how Japanese governmentality has shifted by being confronted by and drawing on changing "Others" in the late twentieth and early twenty-first century. In discussing this shift, the paper draws in particular on Sheldon Garon's (1997) analysis of Japanese state governmental strategies, and concludes that sports mega-events make for a particularly uncomfortable and high-profile reminder of the impossibility of the Japanese state's current assumptions regarding how an ageing society can continue to resist becoming more integrated in global population flows.

Japan and the outside

Since the period of post-war reconstruction, Japan has opened up economically but to a much lesser extent, demographically. At the same time, the *gaikokujin* ("outsider") is a significant specter haunting debates about security and safety. Now, in the long period of decline following the collapse of the bubble economy of the 1980s, intensifying economic globalization and the rise of the "BRIC" nations (Brazil, Russia, India China), but in particular the increasing regional influence of China (see Yee-Kuang 2010), and with the most intense demographic ageing of any national population, Japan's political elite are struggling to cement Japan's place in the world. This struggle has been most visible in the longstanding attempts to lobby for a permanent seat on the UN Security Council (Drifte 2000) and other softer power strategies to secure a global political status equivalent to the nation's still strong, but now already declining economic influence.

Part of this international engagement entails hosting major global events, including sports mega-events (Collins 2007). However, it is in the nature of global events that they involve an opening up to the outside, and therefore remain beset with a particular anxiety over foreign threats to order. In Japanese public discourse at least, such outside threats often gloss over internal dynamics, at the same time as presenting a welcoming face to outsiders and the global media. This could be explained by reference to a simplistic form of history and the supposedly unchanging cultural characteristic of xenophobia as earlier western analysts once tried to show. However, as we shall show, this kind of orientalist explanation is difficult to sustain when one looks back on the preparation for earlier mega-events, both sporting and non-sporting in Japan. The nature of the "Others" seen to be threatening order has not remained static in the post-WW2 period but has shifted. A final complicating factor is that, as in many other countries, the "War on Terror" has provided a justification for multiple existing contradictions to coexist and molded a makeshift single discourse of threat without resolving the underlying problems.

This paper pulls these threads together through an examination of a chain of mega-events, from the Tokyo Olympics of 1964, through the Sapporo Winter

Olympics of 1970, to the Japan/Korea World Cup 2002 and the G8 summit in Hokkaido in 2008, and the public policy developments before them and between them. In particular, it examines the role of the politics of nationalism, and the generation of fear, in the context of the "one-world" ideology represented by mega-events. These different mega-events both represent and in some cases intensify the dominant politics of the era.

Whereas Sandra Collins (2007) has argued that the Japanese state has used sports mega-events (as well as the export of its baseball stars to the US Major League) for international advantage, in contrast we argue that such sports and other mega-events in Japan have represented an opportunity for internal social mobilization by the state by generating the image of an enemy. This mobilization is used to underpin a discourse of advancement and technological modernization but which also reveals the failure of earlier forms of indigenous surveillant governmentality based on moral suasion (Garon 1997), and its replacement by a new global surveillant governmentality based on risk and technological control.

The prominent Japanese sociologist Masachi Ohsawa (1998; 2008) divided post-war Japanese governmental/ideological history into three periods; the Era of Ideals (1945–1970), the Era of Fiction (1970–1995), and the Era of Impossibility (1995–present). We will utilize this conceptual framework to inform our theoretical consideration of the socio-political significance of Others for Japanese society in relation to mega-events.

During the Era of Ideals, Japan concentrated on its reconstruction from the damage caused by its military defeat. In the immediate post-war period in Japan, the Other was regarded mainly as the West, with which Japan sought to identify. This was essentially a recreation of the official modernization discourse that had been dominant in Japan from the Meiji Restoration of the late nineteenth century until the period of national exceptionalism which grew in the 1920s and the imperialist expansion from the 1930s. Copying and catching up to the West were the nation's main objectives. The mega-events that were held during this period were: the Tokyo Olympic Games of 1964, and the Osaka Expo of 1970. Successfully hosting those international events by welcoming such Others, particularly Westerners, to Japan involved the calculated performance of national reconstruction and the renewed prosperity of Japan.

During the Era of Fiction, Japan became one of the wealthiest nations in the world, at least on paper. Based on its notional economic power, Japan seemed to be more self-confident and even arrogant in its relationship with not only Asian nations but also the same Westerners—the European countries and its former occupying power and global "sponsor," the USA—that it had previously sought to impress. As is well known, an increasingly fierce economic conflict arose between Japan and the USA during the 1980s, and in the USA the renewed "threat" of Japan was emphasized in politics and culture. For its part, in this period, the Japanese image of the West as Japan's positive Other shifted to an increasingly negative position. Ultra-right politicians like Ishihara Shintaro (at the time of writing, a longstanding Mayor of Tokyo) openly declared his anti-US attitude with regard to the economic conflict with the USA, co-authoring the popular polemic, *The Japan that Can Say No* (Morita and Ishihara, 1989) and extending this briefly into a pan-Asian anti-American front, with President Mahathir of Malaysia. During the Era of Fiction, Japan's relationship with the West changed so that the West was no longer the ideal with which post-war Japan should identify, but the rival with whom Japan can economically compete. In the context of Japan's economic growth and the contrasting economic crisis in USA, the mega-events held in Japan in this period, for example the Tsukuba International Science Expo of 1985, seems to have sought to demonstrate not Japan's *acceptability* but its global economic and technological *superiority*. Whilst this might not have been true of the mega-event right at the start of this period, the Sapporo Winter Olympics of 1972, it too still reflected significant changes (see below).

Ohsawa argues that the Era of Fiction came to an end with the failure of the state when confronted with the double shock of the Great Hanshin Earthquake and the terrorist gas attack on the Tokyo subway system by the apocalyptic cult, Aum Shinrikyo, in 1995. Fictions of Japan that had already been damaged by the end of the bubble economy in the late 1980s could no longer be sustained. During the ensuing Era of Impossibility, Japan has been faced with an ongoing and very serious economic recession, a crisis of an ageing population which the state cannot afford and the impact of economic globalization and labor mobility. As the latter has became more prominent in Japanese society, the image of the Other as an outside threat to the traditional way of life and business that Japanese people have been accustomed to, has become more terrifying for the public at large. As Japanese society went into profound crisis in the 1990s, the fear and anxiety over foreign Others has drastically risen and political discourses that fostered the nationalistic and xenophobic mentality seemed to pervade among the public at large. Characteristic mega-events held in this period were the FIFA World (Cup Korea/Japan in 2002), the Aichi Expo of 2005 (Murakami Wood and Abe forthcoming) and the G8 Summit of 2008. Each event showed the distinctive sociopolitical significance in Japan's welcoming Others from abroad and at the same time confessing its fear of unwelcome Others who were expected to threaten the safety and security of Japanese society. In each event, we can observe a widespread tendency among Japanese people towards an emotional fear of Others stoked by media and politicians, even though such a fear was unreasonable and lacked real content.

We will now consider these two events from the current Era of Impossibility in more detail, while drawing on comparisons with events from the two previous eras.

The 2002 FIFA World Cup

One of us has previously analyzed the 2002 FIFA World Cup in the context of media relationships with state and citizens (Abe 2004). However, the World Cup was important in several other ways for our argument here. The first was that it was jointly hosted between South Korea and Japan. This had a strong symbolic significance as Japan had been the colonial occupier of the Korean peninsula until the end of the Pacific War. Now they had an equal relationship, but it was of course much less simple than that. Since the end of WW2, both Japan and South Korea had become subordinate powers in a strategic relationship with the USA. Japan was occupied by the USA from 1946 until 1952 and the US military maintained—and despite the official occupation being long past, continues to maintain—its regional bases in the Japanese archipelago. Korea was divided into two in a war fought between the USA, and its allies, and China. Japan's rapid economic development can be traced partly from its position as a manufacturer and supplier of military materiel for the Korean War, and it was the investment pumped into South Korea by the US alliance after the end of the war that led to its own rapid economic development (Stubbs 1999). They also now share a common "enemy": the People's Republic of Korea (North Korea).

The history of the relationship between Korea and Japan remains highly controversial in both countries. Therefore, for some people, co-hosting the 2002 World Cup was a chance to pave a new way forward, a new "future-oriented" (and correspondingly, forgetful) relationship between Korea and Japan. In the context of this socio-political relationship, the significance of Koreans as Others was very ambivalent in that the World Cup fostered genuine mutual understanding between Korea and Japan at the formal level (for example the official ceremony and mainstream mass media discourses), but at the same time nationalistic and racist discourses concerning Korea could still be observed on the Web (Yamano 2005). With respect to the fear of Others coming from abroad, the image of the hooligan, which was constructed almost entirely by the police through the media, was typical in that those images seemed not to be based on the objective estimation of the how risky those foreigners were, but on emotive images, ironically, largely drawn some time past.

Previous sports mega-events

The second factor is that, of course, the 2002 World Cup was far from the first sports mega-event to be held in Japan. Japan was the first nation with a majority non-white population to host the Olympics. Japan hosted two of the first three Olympic events to be held outside Europe and North America: the Games of the XVIII Olympiad in Tokyo in 1964, and the XI Olympic Winter Games in Sapporo, Hokkaido in 1972 (the Games of the XVI Olympiad in Melbourne in 1956 was the other), both of which took place in the Era of Ideals. The Tokyo Olympics can be seen as one of Japan's "rewards" for its economic expansion and support for US interests in the Korean War and beyond. It also represented the symbolic end of a period of "punishment" by the Olympic movement, which had removed the 1940 summer games from Japan after it launched its colonial invasion of Manchuria. In recent years, Japan has looked back on the period after the end of the occupation until the Olympics as a special period when a supposed rediscovery of Japanese communal and social values combined with growing prosperity resulting from hard work. This "Showa nostalgia" is justified to some extent in that the Olympics also represented a culmination of a process of rebuilding and the representation of a newly confident but non-aggressive Japan to the world.

The Olympics themselves were judged a remarkable success. The flamboyance of the stadia architecture, particularly the Yoyogi complex designed by Tange Kenzo, and the opening up of a country still little known by the rest of the world, except for the impressions formed by the cruelty of war, all added to what was perhaps the last Olympic Games to be free of controversy, adding to the impression of a more "innocent era." However, this impression of innocence is also a resource for contemporary official security discourse in Japan, which makes frequent reference to rising crime, particularly crime by foreigners, even though the crime rate during the 1950s was very high in comparison with today (Hara et al. 2001). But the combination of deliberate contemporary official myopia and popular media-abetted nostalgia is very powerful.

During this optimistic era, deeper political economic forces were at work. The Tokyo Olympics in retrospect sat on the cusp of a change in both Japanese development and in global capitalism. The Olympics saw the beginning of the end of what Eric Hobsbaum (1994) characterized as the "golden age of capitalism," a period which might be more accurately described as a truce between the interests of capital and society in capitalist countries, during which social institutions were built up at the same time as economic development increased rapidly. However, this "truce" broke down with the arrival of a new kind of voracious and destructive capitalism, which transformed major cities, and particularly Tokyo, with bewildering speed, and left the newly reformed "neo-traditional" cityscapes of the post-war reconstruction a distant memory (Jinnai 1995; Wood and Abe, forthcoming).

By the time of the 1972 Winter Olympics in Sapporo, just eight years later, it was a different Japan and a different world. The post-war consensus had shattered and the promises of the consumer society were being challenged by the expectations of a new generation raised after the war. The promise of multi-party democracy, in which anything seemed possible after the occupation was over (there was even a Socialist government with a Christian Prime Minister for a brief period) had been replaced by the grey right-wing hierarchy of the Liberal Democratic Party, with the only points of contention being patronage and internal factionalism. Like Europe and the USA, Japan had been riven by social unrest from the universities to the factories. This unrest came from both left and right and beyond. On the left were both progressive socialist movements and also more extreme and violent terrorist organizations, in particular the peculiar Japan Red Army Faction. On the right were both conservative traditionalists and neo-fascist militarist organizations, like the (equally peculiar) Tatenokai, a militia formed by the writer and activist, Mishima Yukio (Apter and Sawa 1984; Katzenstein and Tsujinaka 1991; Katzenstein, 2008).

At the same time, conflict in the Middle East between Israel and displaced Palestinians had intensified with violent resistance and, for the first time, international terrorism was used as a tactic by Palestinian para-state organizations. Terrorism did not reach the Olympic agenda until the attacks on the summer games in Munich later that year which then kick-started an ever-intensifying obsession with security at the Games. But the 1972 Sapporo Games had a different feeling than the 1964 summer Olympiad, where it was the radical left and the students who were seen as the main "threat" by the state, not the radical right as in 1964. The games fostered an image of "the enemy" as a left-wing radical, prompting fears of social revolution, both of which were used to justify crackdowns on left-wing groups (Abe forthcoming).

According to the documentary report published by the police office at the district of Hokkaido (Hokkaido Police Office 1972), where Sapporo is located, the expected main threat and target that should be policed were the radical left-wing student movements and the potential backlash from right-wing nationalist opponents fighting against the former. However, the report tells us that the final result of the expected threats was not so damaging at all. There were no serious incidents during the Sapporo Winter Olympic Games. While the reports proudly declared the success of police activities and the collaboration of local governments and the publics with the police, it seems unclear to what extent the "expected threat" of the radical student movements was ever likely. A threat was fabricated and the fact that it did not materialize was seen as a victory for "law and order" not as a failure of intelligence, or a deliberate exaggeration of the likelihood of such a threat in the first place. This theme could be seen writ even larger at the FIFA World Cup much later in 2002.

The foreign threat

There was little in either previous event, however, that characterized the "foreigner" as the enemy. Yet "foreigners" were already part of Japan's social landscape by this time. Koreans, mainland Chinese and Formosans (Taiwanese) had arrived as forced factory-labor during the colonial period and more came to take advantage of the reconstruction under the US occupation (Weiner 1997). They were never allowed to forget their marginal status. Special identification procedures and tokens (including fingerprinting and a special ID card) marked out a second-rate status. Schooling and often housing was separated. What was particularly strange, however, was how this discourse of racial origins developed during and after the Korean War. Families from what then became North Korea were allowed to "affiliate" with North Korean organizations and school their children in North Korean schools with the *juche* curriculum of the totalitarian North Korean state, even though none of them had ever been to "North Korea" and that country was regarded as Japan's enemy. The separate North and South Korean institutions remain today. In fact the North Korean schools are particular holdouts despite increasing state pressure on them, whilst South Korean populations have either assimilated or assert their cultural identity whilst going through the Japanese educational system (Tai 2009).

Thirty years later, however, the foreign threat was paramount when it came to security. Once again, it was not about terrorism. Although the 2002 FIFA World Cup took place after the September 11, 2001 terrorist attacks (hereafter 9/11), its preparation and planning were a product of the pre-9/11 era. In fact, state thinking was the product of even earlier periods and its fears were built on "facts" that were already out of date by the time they were reported. As Abe (2004) notes, the construction of the image of the enemy by a combined government, police and media campaign had two main themes, both of which were based on fear of foreigners: first, the threat from the "football hooligan"; and secondly, a flood of illegal vendors. Each of these had their own characteristics however, and drew on slightly different imagery and lineages. As Abe (2004: 224–25) wrote of the media emphasis on hooligans:

> the basic storyline was almost always the same. These told us how dangerous and violent the hooligans were and warned the Japanese people to carefully prepare against the coming invasion. Even though not many people have actually ever witnessed hooliganism, the image of "brutal, violent hooligans" proliferated in Japan through the media's reporting on the issue.

In this case, it was the violence that was stressed. These hooligans were portrayed as being mainly white Europeans and especially British who threatened a peaceable Japan (see e.g. Scanlon 2002). These stories drew on a legacy of real hooliganism the reality of which was already largely a thing of the past in Europe. Hooliganism had reached its peak in the 1980s and thanks to a combination of the renewed popularity of football, changes to the socio-economic make-up of football crowds, greater security at football grounds, including CCTV, changes to ticketing, larger numbers of private security, and "intelligence-led" and co-operative policing in the UK and Europe, by the early 2000s it was much rarer (Frosdick and Marsh 2005). Yet the Japanese discourse of hooliganism before the World Cup in 2002 acknowledged none of this.

One hundred and ninety cameras were put up in the host cities (Goold 2002; Murakami Wood et al. 2007), and despite the lack of any actual hooliganism during the World Cup—the only disorder was caused by Japanese fans carried away by the unexpected success of their own team—the cameras remained as the main legacy of the event. It is also the case that, as in Europe, intelligence-sharing between police forces in the UK, Europe and Japan, rather than CCTV on Japan's city streets, was the main way in which known troublemakers were prevented from traveling or entering Japan. In the same year, the Tokyo Metropolitan Government installed 50 CCTV cameras in Kabukicho, the major red-light district in Shinjuku ward, in response to an alleged increase in violence around Chinese gangs, and then expanded this to four other areas: Ikebukuro, Shibuya, Roppongi and Ueno (see Murakami Wood et al. 2007 for a preliminary analysis and Murakami Wood (forthcoming) for more comprehensive assessment).

Illegal traders were a different category of foreigners and produced a different tactic. Since the so-called bubble economy of the 1980s, when Japan underwent a second wave of fast economic growth and came to dominate global banking and financial flows for a short period, Japan had seen a new wave of immigration of people looking for work in Japan's factories and service industries. New immigrants came from South America (particularly returning Japanese Brazilians), mainland Asia, and for the first time, from India, the Middle East and Africa (see Clammer 2001). The post-bubble period, whilst it provided fewer long-term jobs, coincided with a long-developing change in Japan's demographic structure, with a rapidly ageing population, and also with a casualization of labor in many sectors, leading to the kind of low-paid, insecure employment, whose risks poor migrants are prepared (or forced) to tolerate. In this context, the state secretly installed face-recognition technologies at the two principal international airports, Tokyo's Narita and Osaka's Kansai (Asahi Shimbun, 30 December 2002: 22). Although it detected no threats, the face-recognition system was rather more interesting as part of a discursive construction of racial difference: really, it was about whose face "didn't fit" in Japan.

Abe (2004) has argued that the cozy consensus of media, state and public violated a normative expectation of the role of the media in a democracy that, according to Habermasian communicative theory, should exist in a kind of productive tension with authority. The lack of opposition or even discussion from the public "shows how very tolerant public opinion is of the use of highly advanced surveillance systems to enhance public security . . . [and] how easily government can introduce a new surveillance system, bypassing any rational discussion about it" (2004: 226). However, it is important not to assume that a media-constructed threat is necessarily the underlying reason for any development. As Norris and Armstrong (1999) showed with the development of CCTV in the UK (and this was stressed again by Coaffee et al. 2009), whilst hooliganism, terrorism and crime against children combined to assure popular support for (or at least indifference to) the introduction of CCTV, the underlying drivers for its introduction and its expansion were political economic ones, firstly in terms of the liberalization of planning policy in the 1980s leading to the decline of urban centers faced with the competition of "safe" US-style out-of-town shopping centers, and later the ready availability of central state funding for CCTV (Webster 2004). Here we can see some discrepancy between the persuasive logic of a media discourse that stresses the socio-political threat to come and the enhancing logic that has a close relationship with the political-economic interests and objectives of state, police and corporations.

The traveling circus of new global governance

In interviews conducted with a number of different police officers at different levels of seniority in 2009, several officers stressed that the World Cup was not really the most important underlying reason for the introduction of the new security and surveillance systems around the 2002 World Cup. In an interview conducted by one of us with representatives of the Tokyo Metropolitan Police

Department in August 2009, a senior officer made the surprising claim (given the previous narrative related here) that the introduction of CCTV in Japan had very little to do with the World Cup and was the product of "the experiences of the G8." When pressed on this view, he specifically mentioned the Gleneagles summit, which took place in Scotland in 2005. The Gleneagles 31st G8 Summit was of course overshadowed by the July 7, 2005 (7/7) terrorist attacks in London, but other than being an oblique reference to terrorism as a general concern, and given that state CCTV was introduced in Japan before this meeting, it seems hard to know what to make of this. Certainly discussions on security in the G8 have been noted as leading to a change in thinking in the policing of participating nations—see in particular the work of Statewatch (2004) on the 2004 Sea Island 30th G8 summit and its influence on biometric borders. Indeed this influence was strongly in evidence in other interviews conducted as part of the same research trip, particularly with the Prime Minister's special IT Strategic Headquarters in regard to the introduction of new central state information systems and ID cards.

Japan had hosted a G8 summit in 2000 (the 26th, in the southern island of Okinawa, still partly occupied by US forces), and it did so again in 2008, with the 34th Summit in the resort area of Toyako in Hokkaido, the most northerly main island of the Japanese archipelago. Security was tight and according to personal contacts in the city at the time, foreign visitors reported being stopped, questioned and searched on several occasions as paranoia about disruptive foreign protests ran high. The assumption that Japanese social movements might have something to say seemed somehow inconceivable to the authorities at this point.

The Hokkaido G8 Summit can be regarded as a special example in that the security issue was so paramount in the context of the global security regime constructed after the 9/11 terrorist attacks in 2001. One characteristic of the Hokkaido Summit was that the image of the anti-globalization movement that was produced by the media discourse was very dangerous and violent and designed to be (it seemed) deliberately confused with a nebulous idea of "terrorism." The images of the anti-globalization movement that were depicted by Japanese media were highly violent and undoubtedly negative. Those images percolated out to the Japanese public through the media coverage of the protest movements against the G8 summits in Genoa and Seattle. As a result, the meaning of anti-globalization movements was interpreted by the Japanese public not through its own political message but through violent and terroristic images. Therefore, it appeared to be reasonable that the public at large accepted the tightening of security by the police hosting the G8 Summit, even if the surveillance policy seemed to be excessive.

While the basic image of members of anti-globalization movements is that of the "external enemy"—the external Others who are coming to Japan from abroad—the image of the "enemy within" was represented through the media discourse on domestic movements in Japan that sympathized with and showed solidarity towards the international anti-globalization movement. Here we can point out the socio-political mechanism that utilizes the negative image and public fear of an "external enemy" to demonize and control the people and organization that are hostile to the Japanese government, labeling them as a potential "enemy within." It appears that some Japanese people were seen to have become almost infected with "foreignness" which manifest itself as violent protest, and by implication terrorism. The irony of this should not be lost on anyone, given that, thus far, the only terrorism to have been perpetrated on Japanese soil has been by Japanese groups (the Japan Red Army already mentioned, and Aum Shinrikyo). Violence has long been a characteristic of extreme Japanese state and non-state political actors.

The G8 was notable too for its deployment of "volunteers"—a misuse of the term, as the volunteers were mostly schoolchildren pressed into service for the event. The G8 was of course a different type of event to the World Cup. The general public was not invited as paying guests or viewers. Therefore, a more generalized kind of surveillance could operate that was more exclusionary, operating as an exceptional space where sovereign power predominated (cf. Agamben, 2005). It was not about sorting the foreigners into dangerous and non-dangerous categories (English hooligans, Chinese and African traders) but really about regarding *all* (non-government) foreigners as suspicious and potential security threats, and contrasted with the supportive civic-minded Japanese volunteers.

Discussion

The socio-political significance of Others for post-war Japan has changed, depending on the economic and political status of Japan in the world. To understand Japan's complicated relationship with Others, it is useful to utilize the contrasting concepts of xenophobia and xenophilia. On the one hand, post-war Japan has showed its apparent "philia" for the Other as the West. Imaginarily identifying with the West was the national desire of the immediate post-war period in Japan, the Era of Ideals. However, in the process of Japan's economic growth and concomitant growing political power of Japan, the relationship with the West caused the economic and political tension between Japan and its ideal Other. As the result, during the Era of Fiction, Japan's relationship with the West became more ambivalent. Finally, in the Era of Impossibility of the post-bubble economy, Japanese society became more inward-oriented. Meanwhile, the fear of outsiders who were regarded as threatening the socio-cultural stability of Japan has risen drastically.

Although it is dangerous to indulge in psychoanalytical analogy when discussing socio-political matters, one can argue that on the one hand, the concept of complex (inferiority, superiority, Oedipus, etc.) is suitable in describing Japan's relationship with Others in the first two eras. On the other hand, the concept of paranoia might be useful in analyzing the socio-political mechanism that brings about the nationalistic, emotional and exclusionary treatment of Others, which is reappearing in the current Era of Impossibility.

We say "reappearing" here, because as we noted in a previous paper, "in a society which has had a very strong sense of coherent identity, now in some degree of turmoil because of growing visible ethnic minorities, fear of outsiders remains pervasive" (Murakami Wood et al. 2007: 558). One can see this most from the feudal suppression of Christians through women's groups' fear of the "pollution of the blood" by US soldiers after WW2 to the fear of hooligans at the World Cup. However, the discussion of these events reveals that the authoritarian response is not so much a standard xenophobia but an anxiety over the loss of control of all kinds. Foreigners have represented this at particular periods in the past and seem to have returned to this role now. However, the anxiety that these events generate comes from many other sources including internal radicalism and simple non-conforming actions. Essentially this is a fear of Otherness more generally, which is believed to violate the socio-cultural conformity of Japanese society. Therefore, while we have to be cautious not to be trapped by cultural essentialism, the conception of cultural conformity in Japanese society remains useful in comparative research on globalized surveillance (Abe 2009).

This is vital when one considers Japanese governmentality. There is a perception that many foreign mid-twentieth century writers were stereotyping when they wrote of the yielding, compliant character of Japanese people. However, they simply did not dig deep enough into what they were seeing. Successive state forms in Japan have sought to create a compliant populace through various forms of surveillant control. Saito (2004) adopted a neo-Foucauldian analysis to argue that this was a form of self-surveillance, but it was the US historian Sheldon Garon (1997) who has captured this most eloquently in his notion of "moral suasion," the persistent and pervasive promotion of a particular kind of social good by the state and its agents. One might therefore define the most recent Japanese form of surveillance society, as a suasive surveillance society or, to adapt John Law (1993), a suasive mode of ordering.

This suasive mode of ordering was already one of the fictions of the post-1970 model of governmentality, but in the Era of Impossibility, anxiety arises from recognizing that suasive surveillance works less and less well in the neo-liberal world order. Young people become "freeters" (casual workers unaffiliated to a particular employer) or even NEETs (Not in Employment, Education or Training). Foreigners who are not responsive to these messages and their style, arrive, and now they want to stay. Even the criminals are not the predictable, pliable Japanese Yakuza but the unpredictable and devious Chinese Triads. In a sense, these global events are merely the harbingers of the new global governmentality, the infiltration of globalization into particular places. Their security can be seen as conveniently exceptional and specific. But the underlying challenge they represent for an old Japanese suasive mode of order remains when they are gone, which is why the cameras stay, which is why the tactical concentration on "foreign crime" remains (despite Japan's very low place in the international league tables on crime).

Partly at least, this is because the ongoing fiction of the success of suasive surveillance only worked insofar as it was tied into the economic success of Japan, or rather the less pleasant aspects of such a directive hand on society would always be ignored so long as economic benefits were likely. The other side of Japan's undesired opening to the neo-liberal global economy has been an increasing gap between rich and poor, and the increasing efforts by the rich to separate themselves socially and spatially, to become what one interviewee called "new samurai," a class that is able to wall themselves off from the rest, as did the traditional design of samurai estates. The myth of egalitarian Japan is now dying alongside the myth of attitudinal uniformity and *nihonjinron* (theories of Japanese uniqueness).

However, the process of changing the mode of order is far from simple and one-way. On the one hand, we can see an emergence of the new global governmentality, which seems to be contradictory to the previous suasive mode of surveillance in Japan. But on the other hand, under the socio-economic process of an increasing gap between rich and poor, the traditional mode of suasive surveillance still persists, not at a societal level, but *within* the increasingly separated social classes. Inside the world where the community of conformity that is inhabited by the rich is relatively vital, the mode of suasive surveillance functions as it always did. Therefore we must pay close attention to not only the transition of the mode of surveillance but also the multi-layered character of modes of surveillance in contemporary Japan, and indeed in other nations and regions.

Conclusion

In the end this paper has been about the mega-event as the interface between global and local or national, as a crucible of Japanese governmental anxieties in Ohsawa's Era of Impossibility. These fears are not appeased by the temporary "island security" of such events, but rather such security signals the early days of a greater upheaval in global government and the emergence of a global form of surveillant governmentality, not imposed on unwilling states, but shared and translated between a transnational ruling class (van der Pijl 1998), of which the "new samurai" are merely the local incarnation, and imposed on their peoples. The new package that arrives with the global economy includes many impossible things: an openness which is in reality the forced smiles of volunteers, freedom that "needs" identity registries, the exposure of the myth of social solidarity in a society that needs ever greater social support for its elderly, and surveillance cameras which do not see anything of consequence. The Japanese state has accepted the package, but the long-term impossibility of Japan's current government ideology has yet to be challenged, and the social mobilization that has accompanied such moves in the past appears to be increasingly aimless and ineffective.

Acknowledgement

This paper contains some references to interviews conducted as part of David Murakami Wood's UK Economic and Social Research Council Fellowship, "Cultures of Urban Surveillance." He would like to thank the ESRC for their support.

References

Abe, K. 2004. "Everyday policing in Japan: Surveillance, media, Government and public opinion." *International Sociology* 19 (2): 215–31.
—— 2009. "The myth of media interactivity: Technology, communications and surveillance in Japan." *Theory, Culture & Society* 26 (2–3): 73–88.
—— forthcoming. "Policing the enemy within: Fear of radical sects and the Winter Games in Sapporo, 1972." In *Olympics and Surveillance*, edited by V. Bajic and W. DeLint. London: Routledge.
Agamben, G. 2005. *States of Exception*. Translated by Kevin Attell. Chicago, IL: University of Chicago Press.
Apter, D. E. and N. Sawa. 1984. *Against the State: Politics and Social Protest in Japan*. Cambridge, MA: Harvard University Press.
Asahi Shimbun. 2002. "Installing face recognition cameras in customs at Narita and Kansai International Airport: Interrogating passengers without them knowing", Hyouji nashini Satsuei: Ryokyakyu o kojintokutei. Narita to Kanku ni 'kao-ninshyo shisutemu'. *The Asahi Shimbun*, December 30, 2002, p. 22.
Clammer, J. 2001. *Japan and Its Others: Globalization, Difference and the Critique of Modernity*. Melbourne: Trans Pacific Press.
Coaffee, J., D. Murakami Wood and P. Rogers. 2009. *The Everyday Resilience of the City: How Cities Respond to Terrorism and Disaster*. Basingstoke UK: Palgrave Macmillan.
Collins, S. 2007. " 'Samurai' politics: Japanese cultural identity in global sport—The Olympic Games as a representational strategy." *International Journal of the History of Sport* 24 (3): 357–74.
Drifte, R. 2000. *Japan's Quest for a Permanent Security Council Seat: A Matter of Pride or Justice?* Basingstoke, UK: Macmillan.
Frosdick, S. and P.E. Marsh. 2005. *Football Hooliganism*. Devon, UK: Willan.
Garon, S. 1997. *Molding Japanese Minds: The State in Everyday Life*. Princeton, NJ: Princeton University Press.
Goold, B. 2002. "CCTV and public area surveillance in Japan" *Hosei Riron (The Journal of Law and Politics, Japan)* 34 (6).
Hara, T., K. Katsura and Y. Tajima. 2001. *Media Kisei to Tero Sensō Houdō (Media Regulation and Coverage of Terrorism and Wars)*. Tokyo: Akashi Shoten.
Hobsbawm, E. 1994. *The Age of Extremes: The Short Twentieth Century, 1914–1991*. London: Michael Joseph.
Hokkaido Police Office. 1972. *Documentary reports of the activities of police at Winter Olympic at Sapporo 1972*, Hokkaido: Headquarters of Hokkaido Police office.
Jinnai, H. 1995. *Tokyo: A Spatial Anthropology*. Berkeley, CA: University of California Press.
Katzenstein, P. J. and Y. Tsujinaka. 1991. *Defending the Japanese State: Structures, Norms and the Political Response to Terrorism and Violent Social Protest in the 1970s and 1980s*. Ithaca, NY: Cornell East Asia Program.
Katzenstein, P. J. 2008. *Rethinking Japanese Security: Internal and External Dimensions*. Abingdon, UK/New York: Routledge.
Law, J. 1993. *Organizing Modernity: Social Ordering and Social Theory*. Oxford: Wiley-Blackwell.
Morita, A. and S. Ishihara. 1989. *The Japan That Can Say "No": The New U.S.–Japan Relations Card*. Tokyo: Kobunsha.
Murakami Wood, D. (unpublished paper). The uneven development of CCTV in Tokyo.

Murakami Wood, D., D. Lyon and K. Abe. 2007. "Surveillance in Urban Japan: A Critical Introduction." *Urban Studies* 44 (3): 551–68.

Murakami Wood, D. and K. Abe. Forthcoming. "The aesthetics of order: Mega events and transformations in Japanese urban governmentality. *Urban Studies*.

Norris, C. and G. Armstrong. 1999. *The Maximum Surveillance Society: The Rise of CCTV*. Oxford: Berg.

Ohsawa, M. 1998. *Sengo no shiso kuhkan (The Space of Ideology in the Postwar Period)*. Tokyo: Chikuma Shinsho.

Ohsawa, M. 2008. *Fuka no sei no jidai (The Era of Impossibility)*. Tokyo: Iwanami Shinsho.

Saito, T. 2004. *Anshin-no-Fasizumu (The Fascism of Safety)*. Tokyo: Iwanami Shyoten.

Scanlon, C. 2002. "Japan prepares for hooligan threat", *BBC News*, April 17. URL: http://news.bbc.co.uk/2/hi/asia-pacific/1935390.stm (accessed June 14, 2010).

Statewatch. 2004. "G8 meeting at Sea Island in Georgia, USA – sets new security objectives for travel." URL: http://www.statewatch.org/news/2004/jun/09g8-bio-docs.htm (accessed June 14, 2010).

Stubbs, R. 1999. "War and economic development: Export-oriented industrialization in East and Southeast Asia." *Comparative Politics* 31 (3): 337–55.

Tai, E. 2009. "Between assimilation and transnationalism: The debate on nationality acquisition among Koreans in Japan." *Social Identities* 15 (5): 609–29.

Van der Pijl, K. 1998. *Transnational Classes and International Relations*. London: Routledge.

Webster, C. W. R. 2004. "The diffusion, regulation and governance of CCTV in the UK." *Surveillance & Society* 2 (2/3): 230–50.

Weiner, M. ed. 1997. *Japan's Minorities: The Illusion of Homogeneity*. London: Routledge.

Yamano, S. 2005. *Manga Ken-Kanrhyu (Cartoon Anti-Korean Boom)* Tokyo: Shinyusha.

Yee-Kuang, H. 2010. "Mirror, mirror on the wall, who is the softest of them all? Evaluating Japanese and Chinese strategies in the 'soft' power competition era." *International Relations of the Asia-Pacific* 10 (2): 275–304.

Chapter 5

"Secure Our Profits!"
The FIFA™ in Germany 2006

Volker Eick

FIFA: a realist "neocommunitarian entrepreneur"

To engage in a meaningful critique of international football tournaments (or, soccer games as some disbelievers say) it seems to me that I need to embed its main shareholder, the *Fédération International de Football Association* (FIFA) and its main product, the World Cup, into current economics, namely into "actually existing neoliberalism" (Brenner and Theodore 2002). As Bob Jessop (2002) has shown, there might be promising reasons for doing so. In the case at hand the path begins in the mid-1970s and starts with João Havelange, a Brazilian business magnate who was then coming into power as the new FIFA President, and who felt himself responsible for turning FIFA into a global business company. He attracted multi-national brands, such as Coca-Cola and Adidas to lucrative sponsorships and sold TV rights, thus transforming the World Cup into big business in terms of global audiences, profits and spectacular surveillance and security (FIFA 2003; Darby 2003; Homburg 2008; Boyle and Haggerty 2009). "In no time, Havelange transformed an administration-oriented institution into a dynamic enterprise brimming with new ideas and the will to see them through" (FIFA 2003: 6). Since "the mid-1970s FIFA has pressed ahead with the commercialization and professionalization of international football. In this way it has systematically extended its financial resources as well as its global radius of action and adopted a new definition of its duties" (Eisenberg 2006: 59).

It is against this background that I analyze the 2006 FIFA World Cup (hereinafter referred to as the World Cup) as an example of the neoliberalization of sports in general and of the securitization of profits by means of surveillance in particular. As has been argued elsewhere (Eisenberg 2006; Eick 2010a), FIFA as a nonprofit-organization is not neoliberal in the strict sense (Brenner and Theodore 2002; Harvey 2005) but its practices conform to what Jessop (2002: 461) has called neocommunitarianism. Neocommunitarianism incorporates the notions of limiting free competition, enhancing the role of the third sector (nonprofits), emphasizing social cohesion, and "fair trade not free trade" in order to serve a "common goal." Based on the understanding of FIFA as a realist neocommunitarian entrepreneur, the purpose of this chapter is twofold. First, it argues for an understanding of FIFA as a neocommunitarian but neoliberalizing organization aiming at profit maximization through its main product, the World Cup. In the second section, the World Cup is taken as an empirical example of how and in which forms urban neoliberalization shapes and is shaped by FIFA. In other words, it addresses the commercialization of (public) space and its hierarchization. The mode of World Cup-production forms and is formed by urban space. The same holds true for the safety, order and security complex. Its strategies and tactics are analyzed in terms of humanware (i.e. people), software and hardware (i.e. technology and law). In essence, FIFA's attempts to implement a security and surveillance assemblage is to be understood as a means to enhance profit.

The FIFA World Cup: a neoliberal "cash machine"

It is striking to what extent FIFA, as the head of the self-declared "football family," has an impact and influence on state and non-state stakeholders before, during and after the World Cup. In the following section, the commercialization and securitization of urban space before, during and after the 2006 World Cup in Germany is analyzed.

Local governments under conditions of global neoliberalization have adopted, among other things, place-marketing, public–private partnerships, and new forms of local boosterism. Consumer attractions such as sports stadia, convention and shopping centers, plus entertainment "in the form of urban spectacles on a temporary or permanent basis have all become much more prominent facets of strategies for urban regeneration" (Harvey 1989: 9). Given the well-known trend of the festivalization of the city since the mid-1980s (Häußermann and Siebel 1993) and given the intensified marketing of the "city as an entrepreneur" since the early 1990s (Duckworth *et al.* 1987), it is the temporal–spatial intensity of the marketing and commodification processes before and during the four weeks of the World Cup that is particularly significant. In marketing its main product, FIFA's "cash machine" (Homburg 2008: 35) makes use of its monopoly. In addition, FIFA's important political role is based on the symbolically loaded meanings that national competitions such as football matches are said to represent (Boyle and Haggerty 2009) and is fueled by FIFA and the host nations alike (Bundesregierung 2006; Brauer and Brauer 2007). The World Cup's importance rests on the "social cohesion" the games are said to be able to provide by amalgamating different classes within a nation state under the rubric of a "national team," or "football family." Further, the games are expected to create a sense of national pride among the populace that should lead to an emotional *welding* of state residents into a *Volk*. By the same token, nation branding is part and parcel of displaying the home country as a superior economic, educational, engineering and enlightened location for business and industry. In 2006, for example, the federal government's official World Cup slogan was "Welcome to Germany, Land of Ideas" (Bundesregierung 2006).

FIFA Spaces: neoliberal commercialization

Mega-events such as the Olympics, World Cups, or even G8 summits are not only high-profile symbolically and emotionally laden happenings, but also key moments of urban entrepreneurialism (Harvey 1989). They are part of an intense inter-urban and national competition that operates on a global scale with direct impacts on the citizenry. Marketing these events now occurs in ways that accord with neoliberal urbanism, the commercialization of urban space, and new crime policies. As Mayer comments (2007: 94): "With so-called mega-events, cities began to engage in subsidizing zero-sum competition.... Packaging and sale of urban place images have become as important as the measures to keep the downtowns and event spaces clean and free of 'undesirables' and 'dangerous elements' (such as youth, homeless, beggars, prostitutes and other potential 'disrupters')."

Different stakeholders immediately took advantage of these new strategies and even pressed for a leading role in "modernizing" the urban environment. While such mega-projects are meant to last, the legacy of mega-sport events is less obvious and follows a unique logic. Even though the Summer Olympics and its little sister, the Winter Olympics, are understood by (critical) scholars and city governments as opportunities for urban renewal (Cochrane *et al.* 1996; Essex and Chalkley 2004; Gold and Gold 2008) and although the Football World Cups serve similar purposes (Pillay and Bass 2008; Black 2010), the International Olympic Committee (IOC) and FIFA only occupy a city for a couple of weeks.

Extending the footprint and the facilities

FIFA extends its footprint by redefining its area of spatial control for advertisement purposes. Until 1998, for example, FIFA's control rights applied only within the precincts of the stadia. In 2002, FIFA extended its sovereignty to the so-called controlled access sites, and since 2006, its realm included the event stadia, the fan miles, "and other official sites" (FIFA 2006b: 39). Thus for more than four weeks, host cities are transformed into oligopolized advertising spaces of FIFA's two sponsoring groups: the 15 "official partners" and the six "national sponsors"—which combined, generated revenues for FIFA worth €752.4 million in 2006 (Pfeil 2005). The steady increase of profit generation over the years also holds true for selling broadcasting, television and filming rights—revenue streams that produced €1 billion (Salz and Steinkirchner 2006).

In line with these commercially driven preconditions of spatial control which must be accepted by all potential host cities, FIFA also asks for "clean" ground-advertising areas by demanding that all spaces claimed to be within its realm are to be policed before and during the tournament. In the case of Germany, 12 stadia, their respective vicinities, the 32 training grounds and hotels, and the 12 official fan miles were specially delineated spaces. Such demands included the "provision of detention rooms," and video cameras "with a zoom facility" to be "installed inside and outside the stadium" (FIFA 2007a: 11). Stadia surroundings and the official fan miles were also to be fenced (FIFA 2009: 6). Furthermore, FIFA's demands extended to "guaranteeing, planning and implementing law and order as well as safety and security in the stadiums and other relevant locations in conjunction with the relevant authorities" (*ibid*.: 10).

FIFA's Official Provider Licenses meant that German beer or sausage were not to be available in the stadia, as beer and fast food rights were held by Anheuser-Busch (Budweiser) and McDonald's. On the 12 official fan miles, the winning sponsors were expected to have the exclusive right to pitch their products when matches were on, and it was only due to protest by the German Organizing Committee (OC), pressured by German companies, that a compromise was found that allowed the sale of selected non-sponsor products (Baasch 2009: 56). Neither the German application form for the World Cup (dating from 2000) nor the FIFA *Pflichtenheft* (functional specifications) of 2003 mentioned the fan miles—and therefore all regulations pertaining to these spaces were negotiated *ad hoc* (Schulke 2007). Immediately after the World Cup, however, when it became clear that the fan miles were an economic success, FIFA announced that it was taking over organization and marketing of these events in the future (Schulke 2007: 51–52), and had already created the trademark Fan Fest (FIFA 2010). Even though the cordoned-off zones around the stadia were called "security rings," their main function was to provide sponsors and the media with exclusive hospitality zones (FIFA 2007b: 121–163). These purported security rings extended, as Klauser (2008) rightly observes, far beyond the stadia and the inner cities. Nonetheless, to reiterate, FIFA (2006a: 2) does not focus on security but on profit: "official training grounds used by the 32 teams must be handed over to the OC free of advertising materials . . . Specifically, fixed perimeter advertising must be covered or removed by the operator of the facility."

To sum up, "for the good of the game" FIFA demands that applicants and host cities have the ability to secure and police FIFA's profits. This is not to say that FIFA is not interested in "peaceful," safe and secure games and uses volunteers, state-funded police, military, and private security guards towards that end. But all this security activity is centered on increasing profit. In the following, I focus on the commercialization of the fan miles and the "hospitality rings" around stadia.

FIFA Spaces: commercialization in the field and in the rings

As shown above, FIFA forces all applicants for hosting the World Cup (the nation state as well as the respective host cities) to accept in advance all branding conditions and commercialization interests laid down in the so-called FIFA Regulations (FIFA 2006a). From 2002 onwards, FIFA itself extended its right to control urban space further, thus seeping even deeper into neoliberalizing cities. In other words, a nonprofit organization, FIFA, opens up a space for the profitability of itself and of some of the world's largest multi-national companies.

According to Klauser (2008: 181), the security rings around all stadia had to be handed to FIFA as "neutralized space" with all pre-existing advertisements removed. Rules and regulations for the commercialization around the stadia not only applied to the official sponsors but, in turn, also against their competitors. As Klauser (2008: 181–182) notes, even local car garages had to remove advertisements and restaurants had to hide their exterior beer signs (to protect Budweiser). In Munich and Hamburg, for instance, cranes were used to remove the huge advertisements for the insurance corporation *Allianz* and the online marketing company *AOL Germany* (Wilson 2006). As FIFA states, the "OC, FIFA and the stadiums/cities have simply created 'controlled areas'—on an individually agreed basis—in the direct vicinity of the stadiums. In these areas no alternative events should take place . . . to ensure the seamless organization of FIFA World Cup fixtures" (FIFA 2006b: 1).

Hence, we cannot only speak about the "wedding" (Homburg 2008: 41) of football, television and the sports industry but about the wedding of commercialization and securitization as well. It is not only sponsors and non-sponsors, the urban and non-urban citizenry who are affected by FIFA's regulations, but also all administrations from the global to the local scale. The influx of FIFA on regulating urban space as a market and commodity already raises concerns about the democratic conditions of the common weal before, during and after this mega-event—as it does in terms of security. It is to this topic the chapter now turns.

The FIFA World Cup: a neo-liberal security gaze

Neoliberalization promotes a shift away from Fordist–Keynesian forms of government to forms of governance more focused on managing and organizing devolved centers and resources, negotiating "both policy and implementation with partners in public, private, and voluntary sectors" (Stoker 2000: 98). Whereas the role of the (nation) state, according to the well-known metaphor of Osborne and Gaebler (1992: 25), has shifted from "rowing" toward "steering," private industry, and, as FIFA shows, nonprofits as well, thereby achieve greater influence. As part of this development, security functions previously regarded as the domain of the state have been privatized and outsourced, and these shifts in governance and the resulting proliferation of market opportunities are closely connected to the growth of the private security industry (Eick 2006). Even more interesting, FIFA does not only set the commercial rules but the preconditions for security settings as well, suggesting that FIFA demands that host cities "row" in different ways.

Delegating, defining and dividing risks

Several scholars (Ericson and Haggerty 1997; Ericson 2007; Boyle and Haggerty 2009) claim that a more market-oriented and individualized view of security means that consumers are to a specific degree more responsible for their own security, both in terms of their behavior and in terms of making provisions for their own protection. The well-noted prevalence of risk-based thinking, of a suspicion-fueling "precautionary logic," and the state's demand to meet its expectation that everyone, at best, should police herself, however, does not at all apply to FIFA. In fact, the opposite is true as all policing forces, all policing technologies, all World Cup-related safety insurances (FIFA 2006a, 2007a, 2007b), in fact all *risks*, so to speak, are to be covered by the state and have thus been *delegated*.

If we are to understand the term "steering" as, among other (intellectual) activities, the practice of *defining*, the "football family" has been able, with the UEFA taking the lead as early as 1982 (Tsoukala 2009: 29), to *define* risks. From the 1990s onwards, bans on alleged hooligans spread all over Europe, creating what has been called "legal vagueness" for football fans (Tsoukala 2009: 105) and backed by a 2006 European Council's decision that a "risk supporter" can be regarded as a person "posing a possible risk to public order" or the possible risk "of anti-social behavior" (cited in Tsoukala 2009: 109). The highly contested German data bank *Gewalttäter Sport* (containing the personal details of sports related violent offenders), introduced in 2000, was even declared unlawful by the Higher Administrative Court in December 2008, but is still operating (Deutscher Bundestag 2009: 2; Deutscher Bundestag 2010: 1–3).

The power to define and delegate risks does not necessarily mean that FIFA is non-cooperative. In times of networked governance, the opposite is true. This is not surprising as FIFA is in line with the strategies (and even part and parcel) of urban and global elites (Sugden and Tomlinson 1998; Lenskyj 2008). Even though host cities *pay* in economic terms, because hosting World Cups is not lucrative for host cities (Bundesregierung 2006), they (aim to) *earn* in terms of "social cohesion" and ideology (Brauer and Brauer 2007).

In addition, the state-led new crime prevention approaches such as redefining cleanliness and order, socio-spatial rather than individual orientations, and the "punitive turn," are shared by FIFA, the hosting cities and states alike. In as much as public inner-city spaces and mass private property are "sanitized" and strictly controlled (Kempa *et al.* 2004; Flint 2006), the stadia and the fan miles are standardized and people who want to use such spaces have to meet "normalized" behavioral standards (Görke and Maroldt 2006; FIFA 2007a). Whereas the "deviant individual" or the "dangerous classes" used to be the state's central regulatory preoccupation until the mid-twentieth century, it is now more the spaces those individuals and groups occupy that are focused on in hopes of controlling troublesome behavior (Belina 2007; Eick 2011a). For example, the homeless in Vancouver learned from the Winter Olympics about the exclusionary and repressive effects of mega-events, orchestrated by private security companies and the police's increased use of "infraction tickets" (OHCHR 2009: 18; Zirin 2010). Measures aimed at "undesirables" such as (foreign) youths, the homeless, prostitutes, or panhandlers (including increasing incarceration rates) were also evident in the 2006 FIFA World Cup (see below), or the 2008 UEFA Cup in Austria and Switzerland (Klauser, this volume).

Such shared attitudes include new forms of cooperation such as "police-private partnerships" (Stober 2000), the mobilization of *civil society* in the form of volunteers (Moreno *et al.* 1999; Emery 2002; Bach 2008: 151–154), and the substitution of humanware by software and hardware (Aas *et al.* 2009), plus an interest in elaborating tactics of "strategic incapacitation" (Noakes and Gillham 2006). All these practices reverberate in the rules and regulations that FIFA (2006b, 2007a) imposes, as the following section shows.

FIFA rules: securitization in the field

FIFA's rules and regulations for behavior standards in football stadia resonate with the redefinition of deviant behavior, the return of *order* as an issue for crime fighting, and the creation and redefinition of new crimes and disorderly behavior by (local) administrations. The zero-tolerance politics deployed worldwide (Smith 1996), the Anti-Social Behaviour Orders in the UK (Flint 2006), the area bans for "undesirables" in Germany (Belina 2007), the policing of the urban poor by rent-a-cops (Eick 2006) and nonprofits (Eick 2003) serve as ample evidence of this trend.

The 2006 FIFA World Cup in Germany saw the largest display of domestic security strength since 1945. During the four weeks of the World Cup in June and July, more than 220,000 police officers from the 16 *Länder*, an additional 30,000 from the Federal Police, an unknown number of secret service officers, 7,000 military guards, and about 18,000 rent-a-cops were deployed (Buhl 2006). In addition, more than 20,000 security-screened citizens took part in security and safety measures as either stewards or volunteers (Görke and Maroldt 2006: 3). A notable exchange of German police officers with colleagues from neighboring countries occurred and more than 500 foreign police were used (Bach 2008: 151), cutting across national borders and, thereby, neglecting any constitutional restraints in the respective countries. Finally, FIFA has its own internal policing entity, the Task Force "For the good of game" which, in 2007, was transformed into a new body (the "Strategic Committee") in order to "to resolve problems within the family, rather than let rulings be made by a judge who comes from outside the world of football" (Platini 2007).

This overall effort was supported and mediated by sophisticated surveillance, information, identification, and communication technologies—the security "hardware" and "software" of the World Cup. These included: 200 data banks containing more than 18 million data files (Averesch 2009: 6); RFID ticketing systems; and video surveillance systems that monitored stadia and their surroundings, hotels hosting the teams, and public viewing zones in the host cities (see below). The police also used Automatic Vehicle Location systems; Sniper Locating Systems; and robots—equipped with video cameras, radar sensors, temperature gauges and infrared scanners to detect potential bombs and explosives in Berlin, Frankfurt, and Leipzig (Eick *et al.* 2007). Tournament venues and their environs as well as public viewing locations in downtown areas were converted into high-security zones with access limited to registered persons and pacified crowds only. Even selling sausages became a security issue. About

150,000 persons who applied for jobs during the tournament were security-screened by secret service officers and the respective computer systems in order to be accredited (Eick *et al.* 2007). In addition, NATO provided airspace surveillance with two Airborne Warning and Control System planes (AWACS) on loan to the German government to control airspace over the host cities. In US military terms this entailed a complete C4ISR system (command, control, communications, computers, intelligence, surveillance, and reconnaissance).

FIFA rules: civil rights and liberties

Even though the 2006 FIFA World Cup was heralded as a total success both by the government and media in terms of organizational capacities and image production (Bundesregierung 2006; Brauer and Brauer 2007), human rights groups and fan clubs raised questions about how the event impacted on civil rights (Eick *et al.* 2007).

The aforementioned data bank *Gewalttäter Sport* was used during the 2006 World Cup to ban people from stadia and public viewing zones and, during the 2008 UEFA Championship in Austria and Switzerland, to ban people from entering those countries. Data were also transferred to the respective police forces (Tsoukala 2009: 111–116), as the number of persons whose data are stored in the German data bank grew from about 6,500 in 2004 to 9,400 in 2006 to 10,711 in 2009 (Deutscher Bundestag 2009: 3; Deutscher Bundestag 2010: 1). According to the German government, 12,149 data files on German hooligans were available before the 2006 World Cup. According to the police, partnering police organizations provided an additional data bank of 9,000 data files on foreign 'hooligans': 8,450 German 'hooligans' were contacted by the police at home or at their workplace; 3,200 local banishments from inner cities, public viewing areas, and stadia were declared; 131 stadia bans were issued on-site and 587 purchased tickets were blocked based on existing stadia bans; an additional 910 notification requirements were issued; and 210 persons were sent into temporary custody. Finally, 370 people were refused entry into Germany (data collected by the author).

In as much as "private logics circulate through public institutional domains" (Sassen 2006: 195), rent-a-cops and volunteers have become part and parcel of the "policing family." Between 900 and 2,300 rent-a-cops, called supervisors and stewards, were deployed per game as were, on average, 1,370 private security officers. An additional 300 volunteers were deployed per game in the security zone (in total about 20,000). The total number of about 18,000 rent-a-cops—coordinated in a consortium that, in effect, regulated competition between the companies in a neocorporatist way—came from 12 different companies (Buhl 2006: 202). All in all, 87,680 security personnel were deployed just to secure the matches.

Whereas it was possible to collect information from the police about their banning practices, we know nothing about the bans issued by their for-profit and nonprofit counterparts (Bundesregierung 2006; Bach 2008) beyond journalistic evidence (Görke and Maroldt 2006: 3). According to Bach (2008: 152–153), the 12,000 security volunteers were divided in sub-groups for information services outside the stadia and to provide information, guidance, safety and security services within the security rings around the stadia (entrance regulation, grandstand control, lost-property office). They were "integrated into a stringent hierarchy, organized in small groups with a team leader" and were "under control of the local FIFA security officer." In addition, FIFA developed guidelines to encourage cooperation between the rent-a-cops (supervisors and stewards) and the volunteers that read: "The security and order staff is authorized to direct the volunteers and, depending on the situation, is allowed to integrate them into the security and order tasks" (cited in Bach 2008: 154). In general, the security responsibilities of the state police and commercial and volunteering security forces were spatially divided. Private security officers were deployed on private space (stadia) and temporarily privatized space (fan miles) in order to provide information, to control entrances, to conduct bodily checks, to coordinate parking lots, and to control tickets (Schmidt 2007: 28). The monopoly on the legitimate use of force, at least officially, remained in the hands of state police who primarily controlled public space and, depending on the host city, the fan miles as well.

In Hamburg, for example, the police controlled the public viewing area at Heiligengeistfeld. In Berlin a private security company controlled the fan mile Straße des 17 Juni, with only a few police officers in the public viewing area, although the police were on constant standby with additional squads in the surrounding areas (Falkner 2006: 12). In addition, the Berlin police held sway with their demands for higher-security measures at the Berlin fan mile which was enclosed by a 5.3 kilometers long and 2.20 meters high fence – with the additional monies for such fencing being paid by the Berlin government (*ibid.*). With regard to the governance of local security, what emerged in and around the fan miles was a security mix of state and private policing bodies both financed with public monies during a commercial event.

Concerning the spatial reorganization of urban space, as Klauser (2008: 178) has shown, fan miles concentrated fans in selected areas in the city, and in doing so classified, separated, symbolically marked, materially arranged, and in essence controlled public space, which extended to harboring specific norms, values and constraints. It is true that most of the fan mile organizers did not ask for an entrance fee, and only Cologne, Munich and six additional non-host cities relied on admission fees (Klauser 2008: 179) and buying drinks was not a precondition to watch the matches. Nevertheless, the claim of the German criminologist Thomas Feltes that "the World Cup has been democratized by public viewing" (2006: 9) is at least confusing if not irritating. On the contrary, what emerges during the World Cups in the host cities is in fact the spatialized suspension of democracy. Of particular public concern was the wide-ranging video surveillance assemblage established during the tournament. For this reason the next section analyzes the devices that were deployed and what remained afterwards in terms of securitized urban space.

FIFA Rules: securitization and video surveillance, before, during and after the World Cup

Compared to countries like the UK or France (Norris and Armstrong 1999; Töpfer and Helten 2005), Germany's video surveillance coverage has only a limited reach into public areas. Due to the "right to informational self-determination," readjusted in 1983 (Töpfer 2007), Germany still legally constrains the use of video surveillance in public space. The right to informational self-determination includes the provision that there should be no surveillance without informed consent except in the "prevailing general interest" and with a clear legal basis to be codified in a federal statute. In addition, according to most state police acts, public area video surveillance shall be limited only to "crime hot spots" (Töpfer 2007: 214). Notwithstanding such ostensible constraints, video surveillance is a reality.

From the 1990s onwards, the World Cup has entailed modernizing police equipment and extensive training in video surveillance-mediated crowd control (Töpfer 2007). In June 2000, when FIFA decided that Germany would host the World Cup, public area video surveillance was operating in only four cities, all of them in Eastern Germany. In May 2005, the National Security Concept was publicized and the Ministry of the Interior called for intensified video surveillance and a (not realized) facial recognition system, but left decisions about implementation and costs to the *Länder* and local authorities. In February 2006, the *Länder* Ministers of the Interior appealed to the respective hosts of public viewing events to deploy video surveillance but did not make it a mandatory condition. In May 2006, video surveillance systems were operating in around 30 cities.

Video surveillance has been the core of the national railway company's (Deutsche Bahn) 3S-system (service, security, cleanliness) since 1994, shortly after the company was privatized. In addition, video surveillance networks operate in most public transport systems of all major cities, some of them modernized in preparation for the World Cup (Töpfer 2007). New video surveillance systems were installed, or existing ones modernized, in all stadia of the First and Second Football League before the World Cup in order to meet the demands of FIFA; the new *Allianz Arena* (opened in 2005) in Munich, for example, is fully equipped with an interconnected RFID and video surveillance system (Beier 2006). The German company Siemens installed the vast majority of video surveillance and entrance control systems in these new or modernized World Cup stadia (Siemens AG 2006).

Permanent video surveillance of public areas was in operation in only 6 of the 12 host cities. Only Hamburg's new video surveillance systems were installed in conjunction with the World Cup, something that occurred in April 2006 (Baasch 2009: 79). Kaiserslautern saw the police roll-out of a temporary extensive public area video surveillance network, consisting of almost 180 cameras (Schmitt 2006: 20), of which 10 remained in public space after the event concluded (Gastauer 2006: 4). According to media reports, Hanover deployed 870 cameras (cited in

Baasch 2009: 158), and all official fan miles were fenced and controlled with video surveillance. But the majority of the 2,000 officially registered public viewing events were organized and secured without video surveillance.

Most video surveillance systems at public viewing areas were dismantled after the World Cup because of legal constraints and costs. At the same time, the World Cup helped to modernize, expand (and centralize) video surveillance systems in major sports stadia, at the railway stations of *Deutsche Bahn* and in urban public transport networks, such as those in Frankfurt, Hanover and Munich. The FIFA World Cup can thus be seen as a catalyst for a neostatist security strategy: with an increasing number of actors being enrolled into the socio-technical network of video surveillance and, in particular, into public transport networks, co-opted by the police. All in all, and given other research on the topic (Boyle and Haggerty 2009), we see the continuation of a trend. World Cups have a catalytic function for video surveillance networks where the interim endures and even extends, such as occurred in Athens after the 2004 Olympics (Samatas 2007).

Conclusions

Mega-events seem to fuel the growing acceptance of and conviction that rent-a-cops (Buhl 2006) and volunteers (Bach 2008: 151–154) are needed to serve security functions. The same seems to be true for the globally active nonprofit FIFA. The growing influence (not necessarily the acceptance) of all three stakeholders, FIFA, rent-a-cops and volunteers, continues a trend that has been witnessed since the early 1990s in Germany as all stakeholders enjoy growing decision-making powers. FIFA is now part and parcel of the neoliberal crusade of global and urban elites.

Nonprofits gain greater influence by merging policy fields such as sports, social welfare, labor-market (re)integration and the local mingling of security, order and safety measures (Eick *et al*. 2004; Eick 2011a). During the World Cup, nonprofits also intensified workfare measures against the long-term unemployed and deployed them as additional security staff (Feltes 2006: 10–11). Finally, private security companies were able to extend their fields of operation due to ongoing outsourcing processes (Eick 2006).

I have argued for an understanding of FIFA as a neocommunitarian realist entity. Backed by Swiss nonprofit legislation, it aims to limit free competition within its realm and is willing to contest public law. FIFA constantly tries to extend and enhance its influence within its community and—at least before and during the World Cups—beyond. It emphasizes its political role, making use of the (nationalized) "social cohesion" function that football ostensibly plays. By thinking globally and acting locally, FIFA has become an important "glocal" stakeholder (Swyngedouw 1997) within sports and fashion, marketing and media, politics and propaganda, even nation-building and nation-branding, and on the ground at the urban level. In addition, the aforementioned Swiss nonprofit legislation makes FIFA tax-exempt worldwide ensuring that it can collect billions of Euros without having to provide any return service to the host nations and cities.

After the event was over, what remained in place were new laws to extend executive power and sophisticated new surveillance technologies. The government of Hesse, for example, amended the Police Laws to allow for the greater use of video surveillance systems in public space, to extend custody for suspects from 48 hours to six days, and to allow for automated number-plate recording and wire-tapping (Nedela 2006: 8). The FIFA World Cups also provide an exceptional experience in comprehensive training for (inter)national security, including military, private security companies, nonprofit organizations, and volunteers. This security gaze has subsequently been transferred to the 2008 UEFA Championship in Austria and Switzerland (Kretschmann 2009; Klauser, this volume), to the 2010 World Cup in South Africa (FIFA 2007a) and to the 2010 Vancouver Winter Olympics (Boyle and Haggerty 2009; Eick 2011b).

Finally, in order to prepare for transnational protest (della Porta *et al.* 2006), experiences from mega sports events have been transferred across the globe through a constantly growing expert network that is reflected in the enduring fragments of designated temporary surveillance networks in different settings (Boyle, this volume). Such networks have contributed to the ongoing militarization of urban public space (Eick *et al.* 2007; Neumann 2007: 3; Führungsstab Bundespolizei 2007: 4).

In summarizing, I want to highlight four points. First, FIFA is a nonprofit organization that shapes and is shaped by "actually existing neoliberalism." Its main business is to market a purported *civil society* activity, playing football, and to transform it into a profitable commodity. In managing and marketing the World Cup, it shapes the social meaning of the game as it is shaped by its main product. Given its nonprofit status, FIFA deploys a particular kind of neoliberalization—a highly regulated and constantly readjusted process containing elements of neocommunitarian thought and practice. Second, from 2002 onwards, FIFA has set the rules and regulations for its mega-events not only within the stadia—from the consistency of the football green, footballers' dress codes, the specifics of security measures, to the commercial logos that can be displayed—but this has now expanded beyond the stadia to the entire host cities and nation. Third, the growing influence of FIFA entails a new mode of governance. The (contractual) relationships between all stakeholders are shaped by FIFA's ability to offer a monopolized product, the World Cup. For this reason the networks created on a four-year cycle are hierarchical and it is a precondition for hosting the games that host nations and cities subject themselves to FIFA's rules and regulation. Fourth, the modes of governance established before, during and after the FIFA World Cups are at the same time *exploited* by other stakeholders, and in particular by the sponsors, the media, and the sports industry. From a state perspective, the World Cup functions as an experimental site where effective measures to police the urban population can be developed. From FIFA's perspective, security just translates into easy but big money.

References

Aas, Katja F., Helene O. Gundhus, and Heidi M. Lomell, eds. 2009. *Technologies of (In)security*. London: Routledge.

Averesch, Sigrid. 2009. "200 Dateien der Polizei erfassen Personendetails." *Berliner Zeitung* (August 15): 6.

Baasch, Stefanie. 2009. *Herstellung von Sicherheit und Produktion von Kontrollräumen im Kontext von Großevents*. Hamburg: Universität Hamburg.

Bach, Stefanie. 2008. *Die Zusammenarbeit von privaten Sicherheitsunternehmen, Polizei und Ordnungsbehörden im Rahmen einer neuen Sicherheitsarchitektur*. Holzkirchen: Felix.

Beier, Andreas. 2006. *RFID – Die Bedeutung der Fortentwicklung neuer Sicherheitstechnologien*. Münster: Universität Münster.

Belina, Bernd. 2007. "From disciplining to dislocation: Area bans in recent urban policing in Germany." *European Urban and Regional Studies* 14 (4): 321–336.

Black, David. 2010. "The ambiguities of development: Implications for 'development through sport.' " *Sport in Society* 13 (1): 121–129.

Boyle, Philip and Kevin D. Haggerty. 2009. "Spectacular security: Mega-events and the security complex." *International Political Sociology* 3 (3): 257–274.

Brauer, Saskia and Gernot Brauer. 2007. *Sport and National Reputation*. URL: http://www.fifa.com/mm/document/afmarketing/marketing/83/31/80/sportverlagenglisch-track-changes.doc (accessed May 13, 2010).

Brenner, Neil and Nik Theodore. 2002. "Cities and the geographies of 'actually existing neoliberalism.' " *Antipode* 34 (3): 349–379.

Buhl, Manfred. 2006. "Ein Erfolgsmodell für künftige Kooperationen von Polizei und Privaten." *Polizei heute* 35 (6): 202–204.

Bundesregierung (eds.). 2006. *Fußball-WM 2006. Abschlussbericht der Bundesregierung*. Berlin: Die Bundesregierung.

Cochrane, Allan, Jamie Peck, and Adam Tickell. 1996. "Manchester plays games: Exploring the local politics of globalization." *Urban Studies* 33 (8): 1319–1336.

Darby, Paul. 2003. "Africa, the FIFA presidency, and the governance of world football: 1974, 1998, and 2002." *Africa Today* 50 (1): 3–24.

della Porta, Donatella, Abby Peterson, and Herbert Reiter, eds. 2006. *The Policing of Transnational Protest*. Aldershot: Ashgate.

Deutscher Bundestag, ed. 2009. *Konsequenzen aus dem Urteil des Oberverwaltungsgerichts Lüneburg zur Rechtswidrigkeit der Verbunddatei Gewalttäter Sport* (Drucksache 16/11934). Berlin: Deutscher Bundestag.

—— 2010. *Sicherheitsmaßnahmen anlässlich der Fußballweltmeisterschaft 2010 in Südafrika* (Drucksache 17/2088). Berlin: Deutscher Bundestag.

Duckworth, Robert P., Robert H. McNulty, and John M. Simmons. 1987. *Die Stadt als Unternehmen*. Stuttgart: Bonn Aktuell.

Eick, Volker. 2003. "New strategies of policing the poor." *Policing & Society* 13 (4): 365–379.

—— 2006. "Preventive urban discipline: Rent-a-cops and the neoliberal glocalization in Germany." *Social Justice* 33 (3): 66–84.

—— 2010a. "A neoliberal sports event? FIFA from the *Estadio Nacional* to the fan mile." *CITY*, 14 (3): 278–297.

—— 2011a. "Policing 'below the state' in Germany: Neocommunitarian soberness and punitive paternalism." *Contemporary Justice Review* 14 (1): 21–41.

—— 2011b. "Shackled and Silenced? From the 2006 FIFA World Cup to the 2010 Vancouver Winter Olympics," in *Rebooting Neoliberalism. Restructuring the urban and beyond*, edited by M. Mayer and J. Künkel, New York: Palgrave Macmillan, forthcoming.

Eick, Volker, Britta Grell, Margit Mayer, and Jens Sambale. 2004. *Nonprofit-Organisationen und die Transformation der lokalen Beschäftigungspolitik*. Münster: Westfälisches Dampfboot.

Eick, Volker, Jens Sambale, and Eric Töpfer, eds. 2007. *Kontrollierte Urbanität. Zur Neoliberalisierung städtischer Sicherheitspolitik*. Bielefeld: transcript.

Eisenberg, Christiane. 2006. "FIFA 1975–2000: The business of a football development organisation." *Historical Social Research* 31 (1): 55–68.

Emery, Paul R. 2002. "Bidding to host a major sports event." *International Journal of Public Sector Management* 15 (4): 316–335.

Ericson, Richard. 2007. *Crime in an Insecure World*. Cambridge: Polity Press.

Ericson, Richard and Kevin D. Haggerty. 1997. *Policing the Risk Society*. Toronto: University of Toronto Press.

Essex, Stephen and Brian Chalkley. 2004. "Mega-sporting events in urban and regional policy: A history of the Winter Olympics." *Planning Perspectives* 19 (2): 201–204.

Falkner, Markus. 2006. "Das Sicherheitskonzept rund um die WM." *Berliner Morgenpost* (July 3): 12.

Feltes, Thomas. 2006. *Zusammenarbeit zwischen privaten Sicherheitsdienstleistern und Polizei bei der FIFA WM 2006*. Köln: VdS.

FIFA eds. 2003. *Fédération Internationale de Football Association. Info Plus: For the Good of the Game*. Zurich: FIFA.

—— 2006a. *Regulations. 2006 FIFA World Cup Germany*. Zurich: FIFA.

—— 2006b. *Marketing FAQs for the 2006 FIFA World Cup*. Zurich: FIFA.

—— 2007a. *Regulations. 2010 FIFA World Cup South Africa*. Zurich: FIFA.

—— 2007b. *Football Stadiums*. Zurich: FIFA.

—— 2009. *Regulations. 2010 FIFA World Cup South Africa*. Zurich: FIFA.

—— 2010. *FIFA Fan Fest for the 2010 FIFA World Cup South Africa. Frequently asked questions*. Zurich: FIFA.

Flint, John. 2006. *Housing, Urban Governance and Anti-Social Behaviour*. Cambridge: Policy Press.

Führungsstab Bundespolizei. 2007. "Deutsche EU-Ratspräsidentschaft und Präsidentschaft der Gruppe der Acht (G8)." *Zeitschrift der Bundespolizei* 34 (2): 4–7.

Gastauer, Christof. 2006. "Videoüberwachung unterstützt Raumschutz." *Polizei Kurier* (Sonderausgabe Wm 2006): 4.

Gold, John and Margaret Gold. 2008. "Olympic cities: Regeneration, city rebranding and changing urban agendas." *Geography Compass* 2 (1): 300–318.

Görke, André and Lorenz Maroldt. 2006. "Tanzen verboten! Eine lateinamerikanische Fiesta in Leipzig." *Der Tagesspiegel* (26 June): 3.

Harvey, David. 1989. "From manageralism to entrepreneurialism: The transformation in urban governance in late capitalism." *Geografiska Annaler* 71B (1): 3–17.

Häußermann, Hartmut and Walter Siebel, eds. 1993. *Festivalisierung der Stadtpolitik: Stadtentwicklung durch große Projekte*. Opladen: Westdeutscher Verlag.

—— 2005. *A Brief History of Neoliberalism*. Oxford: Oxford University Press.

Homburg, Heidrun. 2008. "Financing world football: A business history of the Fédération Internationale de Football Association." *Zeitschrift für Unternehmensgeschichte* 53 (1): 33–69.

Jessop, Bob. 2002. "Liberalism, neoliberalism, and urban governance: A state-theoretical perspective." *Antipode* 34 (3): 452–472.

Kempa, Michael, Philip Stenning, and Jennifer Wood. 2004. "Policing communal spaces: A reconfiguration of the 'mass private property' hypothesis." *British Journal of Criminology* 44 (4): 562–581.

Klauser, Francisco. 2008. "FIFA Land 2006™: Alliances between security politics and business interests for Germany's city network," in *Architectures of Fear*, edited by the Centre of Contemporary Culture of Barcelona, 173–187. CCCB: Barcelona.

Kretschmann, Andrea. 2009. "Governing Emotions. Fußball-Europameisterschaft 2008 in Österreich." *Bürgerrechte & Polizei/CILIP* 92 (1): 65–72.

Lenskyj, Helen. 2008. *Olympic Industry Resistance*. New York: SUNY Press.

Mayer, Margit. 2007. "Contesting the Neoliberalization of Urban Governance," in *Contesting Neoliberalism. Urban Frontiers*, edited by H. Leitner, J. Peck, and E. Sheppard, 90–115. New York: Guilford Press.

Moreno, Ana, Miquel de Moragas, and Raúl Paniagua. 1999. *The Evolution of Volunteers at the Olympic Games*. URL: http://blues.uab.es/olympic.studies/volunteers/moreno.html (accessed May 13, 2010).

Nedela, Norbert. 2006. *Sicherheitsstrategien im Zusammenhang mit der FIFA Fußballweltmeisterschaft 2006*. Berlin: Landespolizeipräsident.

Neumann, Karsten. 2007. *Datenabgleich und Datenschutzinformationen der Einwohner/ Gewerbetreibenden von Heiligendamm und Videoüberwachung im Zusammenhang mit dem G8-Gipfel*. Rostock: KAVALA.

Noakes, John and Patrick F. Gillham. 2006. "Aspects of the 'new penology' in the police response to major political protests in the United States," in *The Policing of Transnational Protest*, edited by D. della Porta, A. Peterson, and H. Reiter, 97–115. Aldershot: Ashgate.

Norris, Clive and Garry Armstrong. 1999. *The Maximum Surveillance Society*. Oxford: Berg.

OHCHR. Office of the United Nations High Commissioner for Human Rights (eds). 2009. *Report of the Special Rapporteur on adequate housing as a component of the right to an adequate standard of living, and on the right to non-discrimination in this context, Raquel Rolnik*. Geneva: OHCHR.

Osborne, David and Ted Gaebler. 1992. *Reinventing Government*. Reading, MA: Addison-Wesley.

Pfeil, Marcus. 2005. "Platz da für die Fifa!" *Die Zeit* 45 (November 4). URL: http:// www.zeit.de/2005/45/FIFA-Republik (accessed May 13, 2010).

Pillay, Udesh and Orli Bass. 2008. "Mega-events as a response to poverty reduction: The 2010 FIFA World Cup and its urban development implications." *Urban Forum* 19 (3): 329–346.

Platini, Michel. 2007. Platini stresses "family values" (October 2). URL: http://www.fifa.com/aboutfifa/federation/bodies/news/newsid=612977.html (accessed May 13, 2010).

Salz, Jürgen and Peter Steinkirchner. 2006. "Der Milliarden-Kick." *WirtschaftsWoche* 23 (June 3): 85–92.

Samatas, Minas. 2007. "Security and surveillance in the Athens 2004 Olympics." *International Criminal Justice Review* 17 (3): 220–238.

Sassen, Saskia. 2006. *Territory, Authority, Rights: From Medieval to Global Assemblages*. Princeton, NJ: Princeton University Press.

Schmidt, Peter. 2007. "Zur Qualifizierung der Sicherheitsdienstleister bei Großveranstaltungen," in *Der Beitrag des Bewachungsgewerbes zur Sicherheit bei Großveranstaltungen*, edited by R. Stober, 27–32. Köln: Heymanns Verlag.

Schmitt, Jürgen. 2006. "Die FIFA-WM 2006 in Kaiserslautern. Eine Nachbetrachtung zu einem gelungenen Polizeieinsatz." *Die Kriminalpolizei* 24 (4): 16–21.

Schulke, Hans-Jürgen. 2007. *Fan und Flaneur: Public Viewing bei der FIFA-Weltmeisterschaft 2006*. Bremen: Institut für Sportwissenschaft.

Siemens AG. 2006. *12 WM-Spielorte und über 100 Projekte im Überblick* (press release, May 15). Munich: Siemens AG.

Smith, Neil. 1996. *The New Urban Frontier*. New York: Routledge.

Stober, Rolf. 2000. "Police-Private-Partnership aus juristischer Sicht." *Die Öffentliche Verwaltung* 7 (53): 261–268.

Stoker, Gerry. 2000. "Urban political science and the challenge of urban governance," in *Debating Governance*, edited by J. Pierre, 91–109. Oxford: Oxford University Press.

Sugden, John and Alan Tomlinson. 1998. *FIFA and the Contest for World Football*. Cambridge: Polity Press.

Swyngedouw, Erik. 1997. "Neither global nor local: 'Glocalisation' and the politics of scale," in *Spaces of Globalization*, edited by K. Cox, 137–166. New York: Guilford.

Töpfer, Eric. 2007. "Entgrenzte Raumkontrolle? Videoüberwachung im Neoliberalismus," in *Kontrollierte Urbanität,* edited by V. Eick, J. Sambale, and E. Töpfer, 193–226. Bielefeld: transcript.

Töpfer, Eric and Frank Helten. 2005. "Marianne und ihre Großen Brüder. Videoüberwachung à la Française." *Bürgerrechte & Polizei/CILIP* 81 (2): 48–55.

Tsoukala, Anastassia. 2009. *Football Hooliganism in Europe. Security and Civil Liberties in the Balance*. New York: Palgrave Macmillan.

Wilson, Bill. 2006. *Stadiums renamed for FIFA sponsors* (June 6). URL: http://news.bbc.co.uk/2/hi/business/4773843.stm (accessed May 13, 2010).

Zirin, Dave. 2010. "As Olympics near, people in Vancouver are dreading Games." (January 25). URL: http://sportsillustrated.cnn.com/2010/writers/dave_zirin/01/25/vancouver/index.html (accessed May 13, 2010).

Chapter 6

Event-driven security policies and spatial control
The 2006 FIFA World Cup

Stefanie Baasch

Introduction

During the last few decades, mega-events have become a keystone of urban development. A necessary condition for a successful event is the security of tourists, visitors, participants and residents. This chapter examines the close links between mega-events, security and urban development.

In what follows I address some of these processes in relation to the FIFA World Cup 2006. These findings are based on a study of media accounts of the World Cup that were present in three different local newspapers and which scrutinized these accounts for themes that included, but were not limited to, the types of dangers and threats that were discussed, the actors that were mentioned, the spatial aspects of danger scenarios and the security measures that were deployed (see Baasch 2009 for details).

Before proceeding to those specifics, some main characteristics of this type of mega-event need to be explained. The World Cup is one of three international mega-events (the other two being the Olympic Games and the EXPO World's Fair), which are defined by their economics and popularity (Schindler and Steib 2005). They are also characterized by their highly structured organizational form, and their close public–private partnerships. However, unlike the Olympic Games or the World's Fair, the World Cup does not occur in a single city but occupies several cities typically within one country. Therefore, the effects on urban development are usually not as high as they are in the context of more stationary events. Mobility between the different host cities also poses an additional challenge during the World Cups. For example, the 2006 World Cup took place in Germany between June 9th and July 9th. During these 25 days, 64 football matches were held in twelve host cities, with Hamburg hosting five matches.

Control societies, mega-events and urban security policies

Over the last thirty years there have been a growing number of studies on the effects of mega-events on their host cities. This research has focused mainly on economic effects and urban planning. Another important component of urban research since the mid-1980s has been concerned with urban security and insecurity. These two fields of research form the background of my study.

This study is primarily based on theoretical ideas about *disciplinary societies* (Foucault 1975) and *control societies* (Deleuze 1990). Much social research on security issues is informed by the work of Michel Foucault, who analyzed the function of disciplinary societies from an historical perspective. In *Discipline and Punish* (1975), Foucault described two techniques which still occupy a central position for surveillance research: enclosure and panopticism. The core principle and technique of disciplinary societies, in Foucauldian terms, consists of disciplining the individual by positioning him/her within different societal institutions (*enclosures*). Foucault linked the enclosure in social spaces to an allocation in physical space. His main example focused on nineteenth-century prisons, where discipline was enforced by surveillance or, to be more precise, made feasible through architectural design, specifically, the *panopticon*. Today, interior and exterior design is still used in a similar way (e.g. as a component of architectures concerned with defensible space), now often supported by surveillance cameras. Not knowing when (and by whom) he/she is being monitored, forces the individual to conform to the written or unwritten rules of social behaviour.

The extent to which Foucault's conclusions on surveillance can be applied to contemporary society is debatable (Haggerty 2009). Furthermore, the diverse interpretations of Foucault's panoptic model have also led to a broad range of conclusions about contemporary society (Simon 2005: 2). Some authors, like Haggerty and Ericson (2000: 607) assert that Foucault's analysis "fails to directly engage contemporary developments in surveillance technology," given his focus on an earlier historical period. Modern techniques of surveillance, which are characterized by the increasing distances between observer and observed, are arguably not comparable to nineteenth-century practices (Giddens 1990). Nevertheless, these points should not lead us to throw "the baby of Foucauldian insight out with the bathwater of an overworked concept of panopticism" (Simon 2005: 2).

In his analysis of control societies Deleuze tried to transfer Foucault's historical analysis of discipline into twentieth-century society. This entailed moving beyond Foucault's analysis of enclosures such as prisons to think about how control now operates in public spaces. According to Deleuze, institutions are no longer exclusively responsible for disciplining subjects, and have merged into the public space where control is permanently exerted. Combined, the insights of Foucault and Deleuze allow me to contextualize social processes which are linked to increased surveillance, the exclusion and displacement of marginalized groups from inner city areas, and the greater self-regulation of people's behaviour. Surveillance facilitates these disciplining processes by fostering a permanent state of emergency. Surveillance is no longer conducted primarily by public authorities, but is distributed between the public and private sectors. This new form of governance is also no longer contingent on public authority, but on the needs of neo-liberal markets. To participate in social life, it is imperative that individuals exercise self-control to meet market requirements. Therefore, control is not only exerted from the outside and regulated by institutions, but it is also active within the individual, who fears being marginalized. Where the disciplinary society sought be inclusive, the control society clearly distinguishes between those included and those excluded.

Foucault and Deleuze offer a broad framework to help us interpret and analyse the cultural and societal processes of control and surveillance. However, they do not offer a comprehensive explanation of how current security needs are being put into practice. In this study, therefore, I investigate how processes of securitization and the strengthening of security policies were discussed and legitimized in the context of the Football World Cup 2006, focusing on the urban domain. I combine the sociological findings of Foucault and Deleuze with perspectives from urban research, mainly focusing on the fields of urban (in-) security and the effects of large events on host cities. The findings confirm that political actors used the dynamics of this mega-event to foster security policies by urban, national and European governments. However, contrary to previous findings and my initial expectations, I found that debates about security were primarily focused at the national and international level, rather than on local security concerns.

Security measures and civil liberties are often seen to be in conflict (e.g. Naucke 2003). Consequently in democratic societies there are recurrent political debates about enhanced security measures which affect constitutional law and restrict civil liberties. Although a full description of the entire field of research which addresses these issues would be beyond the scope of this paper, it is necessary to highlight some basic theoretical approaches to these issues.

Beste (2004: 7), for example, sees security as "a code referring to symptoms of crisis and major structural changes in the post-Fordist city" and characterizes security as a "wild card" because of how it limits public debate on the control of urban space, the construction of public enemies, and the exclusion of dangerous groups. He identifies five factors which have led to a new architecture of social control and new modes of surveillance, especially in urban development:

1 The embrace of entrepreneurial post-Fordist city development which focuses on market mechanisms.
2 Transnationalization and differentiation of policing, which means that policing is no longer exclusively constructed and provided by nation states.
3 Changes in the conception and practice of regulation (especially "community policing").
4 A decrease in police and public budgets in the context of an increasing complexity of urban groups each of which have different needs and interests.
5 Increased levels of risk which have prompted a growth in security markets, especially since 9/11.

Other urban researchers (Smith 1996; MacLeod 2002) identify surveillance and spatialization as resulting from a "revanchist urbanism" in the domain of entrepreneurialism (Harvey 2000). From this point of view, security arguments are used to establish security standards and exclude unwelcomed groups or individuals. I return to this point in the later section on "urban (in)security in Germany."

Concepts from urban research

Hosting a World Cup is similar to other mega-events in that it follows a competitive and costly procedure. So why do cities (or countries) find it desirable to pursue such events? And what are the main outcomes of hosting such an event? In order to assess the influence of large events on urban settings, it was first necessary to review the current state of security within German cities. Here it is useful to link two main fields of inquiry. First is the notion of the "festivalisation of city politics" (Häussermann and Siebel 1993) which focuses on the role played by mega-events and city development, and which has emerged as a key theory in German-language scholarship on urbanism. Second are a number of concepts which have emerged in the field of urban (in)security research. Figure 6.1 graphically depicts these conceptual relationships.

The systematic use of large events by local politicians to push for urban development and remodelling is the central feature of event-driven strategies. Events are closely linked to prospects for increasing a city's attractiveness and enabling it to succeed in international competitions between cities. This, in turn, is seen as a means to derive economic benefits, especially tourism, and to encourage the local population to identify with their home town.

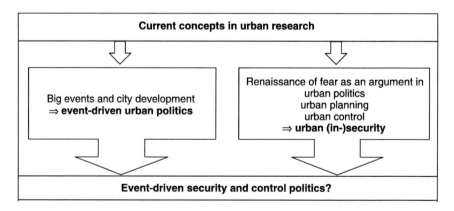

Figure 6.1 Research concepts

Mega-events in current research

The economic outcomes of large events, in reality, often remain below expectations. Surprisingly, this does not seem to have dampened the high hopes officials attach to these events. Even the 2006 World Cup, understood to be a very successful mega-event, hardly had any effects on Germany's GDP or on job creation. Before the event it was estimated that it would produce financial benefits of between €2.5 billion (Rahmann 1999) and €10 billion (Bargel 2005). Several subsequent studies painted a quite different picture of the 2006 World Cup as only having had minor economic effects (Brenke and Wagner 2007). In terms of the job market, the World Cup only produced a small increase in temporary and low-paid jobs (Hagn and Maenning 2007). The effects on infrastructure also turned out to be minimal; an increase of only about 0.02% of the GNP, which can be partially attributed to the fact that the city already had an advanced infrastructure, meaning that there was little need for urban infrastructure programmes (ibid.). This lack of economic impact is not unique to the 2006 World Cup, but is a common feature of other mega-events such as the 1998 FIFA World Cup in France, the UEFA European Cup, and the 2002 Olympic Games in Athens (Matheson 2006: 9; Szymanski 2002: 175f; Samatas 2011). Nevertheless, the 2006 World Cup was one of the largest international media events ever, with 26.3 billion television viewers worldwide watching the games (Schaffrath 2006). This high degree of media attention and the generally favourable reporting are seen as the main cause for the positive opinions that German citizens have about the event.

Table 6.1 presents an overview of the main research results concerning the potential benefits and risks of hosting a mega-event (Baasch 2009; Boyle and Haggerty 2009; Samatas 2008; Pilz 2007; Schindler and Steib 2005; Stott et al. 2006; Simons 2003; Stott and Adang 2003; Lenskyj 2002; Burbank et al. 2001; Ehrenberg and Kruse 2000).

There are a few examples where a mega-event has successfully realized some of its structural promise. The Olympic Games in Munich 1972, for example, are linked to the modernization of the urban infrastructure and development of public transport. The 1986 World Exhibition in Vancouver and the 1992 Olympic Games in Barcelona each produced positive images of the hosting venues and led to a more upbeat marketing dynamic. But the hopes attached to holding a large event often dwindle. Venues can present problems because they cannot be reused, are oversized and expensive, and produce costs which burden the cities' budgets for years.

Still, hosting big events remains attractive for cities and countries because of how they can be used to legitimate other urban developments. The dynamics of the event and the pressure of time accelerate political decision-making. In some cases, the event becomes a focal point for political initiatives which otherwise would only have a slight chance of being realized. Yet another appeal concerns how successful events advance symbolic politics by demonstrating governmental

Table 6.1 Benefits and risks for host cities/countries in the context of big events

	Benefits	Risks
Economic effects	• additional private investments • increased tax revenues due to increased business volumes (esp. dining, tourism, retail)	• follow-up costs assumed by the host city/country • low profit sharing • funds tied up or frozen to finance the event (budget cuts for other projects/plans) • cost explosion (supply shortfall)
Infrastructure	• infrastructure development and upgrading (e.g. public transport)	• oversized and single-use measures (little or no possibility for repeated or continued use) • subordination of other projects
Image	• use of media attention and event tourism for direct marketing • international presentation of venues	• lack of image effects (focus on the event – not on the venue) • negative image (scandals, planning errors) • conflicts between locals and tourists
Organization and urban governance	• participative organization (higher planning reliability and acceptance) • demonstration of organizational capacity • creation of new organizational structures and cooperation	• concentration of power (esp. economic actors) • risk of corruption • loss of control of urban actors (tight deadlines)
Realization	• increase of local identification	• lack of acceptance by citizens • conflicts between locals and tourists • instrumentalization of the event (political purposes etc.) • social marginalization (homeless persons, beggars etc.)
Security and control	• positive influence on the sense of security (social awareness, common identity) • funding of social programmes (fan support) • low-profile strategies (acceptance of police procedures, self-policing of fans)	• event-related security needs (hooliganism etc.) • over-allocation of security • rising level of surveillance • use of event-related dynamics to implement political measures likely not to be accepted by the majority

competence and responsibility. The downside of this is that other political issues tend to be marginalized, such as social problems that are not easy to market (Baasch 2009: 22).

So, while past experiences would advise caution, the expectations for large events remain high. Prior to the 2010 World Cup in South Africa, for example, officials estimated the event would enhance tourism (upward of half a million foreign tourists), provide a R55.7 billion boost to South Africa's gross domestic product (GDP) – including R33 billion spent on stadia and infrastructure – and create 415,000 jobs (Grant Thornton 2008). The South African public seemed to share these optimistic forecasts. A study by the Human Science Research Council (HSRC) reported that 50% of South African respondents expected the World Cup to create jobs and lasting economic benefits, while one-third anticipated personal advantages to accrue from new job opportunities (Tomlinson et al. 2009: 4). These hopes, along with the global media interest in the World Cup, produce enormous pressure on the host country and cities – both from an economic point of view but also in terms of image management (both on a national and international level). In South Africa the symbolic stakes were particularly high given that this, the first football World Cup to be held in Africa, was designed to demonstrate South Africa's ability to manage such a challenging mega-event and advance the country's claim to being a "developed" nation. If the World Cup 2010 is not judged to be a success, it may aggravate South Africa's economic and social situation – and the public will have to pay the bill. And while the majority of South Africans seemed optimistic about such prospects, critical voices were also heard in that country. A few days after the beginning of the tournament, for example, about 3000 people in Durban protested against the massive public investments in World Cup 2010, in light of the fact that 40% of the South African population live on less than $2 a day (Veith 2010).

Urban (in)security in Germany

Debates on urban (in)security focus mainly on topics like access to or exclusion from public and inner city areas, or on the use of public (or publicly available) spaces. Since the mid-1980s "broken-windows theory" (Wilson and Kelling 1982) has been popular in Germany. This theory claims that there is an intimate connection between "disorder" and criminality. Disorder comprises a number of unwanted behavioural patterns, such as littering, or the presence of homeless people or congregations of young people. Urban politics often appeals to "broken-windows" theories to legitimate removing homeless people from city centres in order to make these areas more attractive to middle-class consumers.

There are strong links between event-driven urban politics, the intensification of social-exclusion strategies and increased concerns about criminality. All of these developments are based on the neo-liberal ideal that cities are primarily economic areas, locations for businesses, and spaces for the creation of value. The debates about crime that have occurred since the end of the 1980s are closely connected with the escalating competition between cities, the declining significance of inner cities as commercial centres (provoked by suburban shopping malls), increasing unemployment and the impoverishment of marginalized groups. In this context, increased surveillance capacity (not only by the police, but also by private actors) is seen as a way to encourage self-regulation. Additionally, displacing marginalized people from the inner cities can be understood in the context of "upgrading processes," that enhance security and control standards. In Hamburg, for example, new municipal (2003) and communal (2006) public services were prompted by the increased influence of security and control interests. Additionally, a new law, introduced in 2005, gave the police the power to conduct random identity checks. In 2006, shortly before the World Cup, the police also installed surveillance cameras in the Reeperbahn entertainment district, close to the FIFA Fan Park. These are just a few examples which speak to the increasing levels of control and security in German cities.

Security logics and security actors

Since 9/11, awareness of the vulnerability of (urban) daily life has produced a broad debate on how to prevent terrorist attacks. This has led to enhanced security initiatives on a sub-national level (cf. Graham 2006). All of these urban development processes, combined with concerns about new threats, accelerated and legitimated rising security standards in Germany in advance of the 2006 World Cup.

Security is a key requirement for a successful mega-event – not only at the competition venues themselves, but also in inner-city event areas where public viewings occur. In the case of the Football World Cup, which takes place in several host cities, secure transportation for players, fans and reporters is another challenge. The side effects of such events may turn out to be costly given that the security needs for an exceptional mega-event are quite different than those for routinely proving public safety.

Security for a FIFA World Cup has to be guaranteed by the national government during the application process. This rather remarkable requirement means that national governments must partner with a *de jure* private actor (FIFA), while being responsible for the security of a private event. This, in turn, leads to extensive data exchange between public authorities and FIFA. No one who wants to participate in the event can avoid these data exchanges – whether they are working in the stadium, or as a sportswriter, or simply purchasing a ticket. Such data exchange has already become a standard procedure at large international sports events.

Security was high on the agenda for 2006 World Cup. This included extensive public discussions about terrorism, hooliganism, right-wing extremism, and the types of measures the state would deploy to counter these threats.

Terrorism was understood to be the main threat to the Football World Cup in Germany. The potential for an attack by Islamic terrorists was used politically to legitimize new laws and to expand existing security measures. That said, media analysis (Baasch 2009) reveals that the actors who fuelled public concerns about terrorism also conceded that they had no specific information on imminent terrorist attacks. This illustrates the delicate balancing act of event-driven security politics. On the one hand, risk scenarios are essential for legitimating security measures. On the other hand, these scenarios cannot be so egregious as to curtail tourism.

At one point officials even debated amending Germany's constitution to allow the Bundeswehr (Federal Armed Forces) to engage in domestic policing during the World Cup. Such activities have traditionally been precluded by the German constitution's separation of powers that was established after the Nazi period to protect democracy and prevent coup d'états. The upshot is that the army is only allowed to secure borders or operate outside of German territory under certain conditions (except in cases of disaster). Conservative politicians saw the World Cup as a strategic opportunity to reverse this principle and allow the army to operate domestically. Ultimately, the attempt failed, but the army was asked to cooperate with administrative responsibilities, such as supporting medical services.

The prospect that hooligans might disrupt the World Cup was also debated. Here, it is interesting to note that German and other western European hooligans were seen to pose only a minor threat, whereas hooligans from eastern Europe were labelled as "unknown" threats. The descriptions of these latter groups leaned towards racist stereotypes that reproduced similar images circulating in everyday discourses (Jäger 2006). For example, east European hooligans were described as being more aggressive, and much more dangerous than west European hooligans. At the same time, east European police forces were criticized for not meeting western standards in the fight against hooliganism, for being unorganized and for having fewer standardized protocols for sharing data (Baasch 2009).

The third main fear concerned right-wing extremism. This fear was contextualized in two ways. Firstly, there were attempts by fascist groups to attract media attention by staging events such as marches, which in turn provoked widespread anti-fascist protests. Secondly, right-wing extremism was linked to a controversial debate about whether there were no-go areas in Germany – particularly parts of Eastern Germany – which foreign tourists should avoid.

In the media-driven debates which helped to set the political agenda (Luhmann 1999), only a small core set of unrepresentative political actors tended to dominate the discussion. In Hamburg, these discussions were populated primarily by political actors or representatives of public authorities, especially from the police sector. These spokespersons primarily presented a police-based perspective which fixated on technical security. They rarely discussed concrete issues pertaining to implementing security measures in the venues, nor the local impacts of such initiatives. Alternative understandings of security which focused, for example, on softer measures or on using social work programmes to prevent violence tended not to appear in the media. Most of these debates occurred at the federal policy level, something that tended to marginalize local opinions and actors. In essence, only a small spectrum of opinions on security was voiced in the popular media and these tended to reproduce a police-dominated view of risk and insecurity.

Spatial control strategies

The control techniques employed at the World Cup were politically and spatially varied. The fact that such initiatives operated at diverse political levels reinforces the fact that a mega-event like the World Cup not only influences the host cities or countries but also international politics and relationships.

On the national level, officials sought to prevent threatening people or groups (such as hooligans or terrorists) from entering by securing the border and temporarily reintroducing border controls between the Schengen states. NATO and the Federal Armed Forces established military defence and border controls to avoid potential air or coastal attacks. Border crossings were regulated by travel bans on threatening persons such as hooligans or potential hooligans and by additional inter-country police information exchanges designed to identify suspects within fan groups. New policies were implemented to facilitate information exchange and cooperation between public authorities, the Federal Police and the Federal Armed Forces. While this close cooperation was criticized, mainly in the liberal press, for violating the separation of powers, which is a constitutional principle in Germany (Baasch 2009), these concerns were not sufficient to stop standardized security policies and various forms of international security cooperation from becoming the dominant national strategies in the 2006 World Cup security architecture.

Host cities linked various events to the World Cup. Besides the football matches played in the stadia, the biggest of these side events were the official FIFA Fan

Table 6.2 Control strategies and spatial patterns

- On a transnational level
    - data exchange
    - international police cooperation (e.g. exchange of police officers)
- On a national level (host country)
    - securing borders
    - cooperation between federal authorities
- On an urban level (host city)
    - securing venues
    - surveillance of inner cities/inner city venues
    - displacement of marginalized groups

Parks, followed by other municipally or privately organized public viewings. Most popular during the 2006 World Cup were the so-called "Public Viewing Events" – FIFA Fan Parks for communal viewing of the matches. Such sites were not an entirely new phenomenon, but Germany did set new records in this area with approximately 42% of spectators watching the games from such viewing areas (Geese et al. 2006). In contrast, only 13% of viewers at the 2004 European Football Championship watched from public viewing areas, and that number had only increased to 19% by 2004. This rather marked increase in the popularity of public football viewing can in part be explained by advances in the infrastructure for this type of spectatorship, but also by the good weather during the tournament, the surprising success of the German football team, and the fact that Germany was hosting the event.

During the World Cup, the stadia and the official FIFA Fan Parks were directly regulated by FIFA and subject to the standardized security regulation demanded of all host cities. The Hamburg FIFA Fan Park was built on an inner city event area, which was fenced off and equipped with police video surveillance. In what has become a familiar aspect of mega-events, the surveillance cameras were kept in service after the event and legitimized by their purported value during the World Cup. Another example of attempts to use the event to drive security dynamics was the proposal to introduce a temporary ban on begging. Advocates of the ban argued that it was needed because beggars could potentially damage the city's image and reduce the attractiveness of the inner city for tourists. While this proposal ultimately failed for lack of political support, it again highlights efforts by powerful interests to capitalize on the opportunities provided by the World Cup to initiate control strategies.

Before the event there was considerable debate in the German press about constructing a collection point for people arrested during the event, referred to as the "Hooligan-Knast" (hooligan gaol). Normally, arrested people were brought to local police stations. In cases of mass arrests, such as during demonstrations, arrestees were dispatched throughout the city, due to the limited capacities of the local police stations. This new collecting point would hold up to 150 people and was legitimized as necessary to fight hooliganism and other excesses during the World Cup. It was no secret that this collecting point would also be retained after the event. Fan supporters criticized this measure for criminalizing football fans and undermining the work they had done with fans to prevent hooliganism.

Besides these physical security measures, an additional legacy of the 2006 World Cup was that it normalized increasing security standards. The event created a temporary state of exception which was inextricably linked with dogmatic appeals for the highest possible security levels. Temporarily converting public space into private event areas containing surveillance and access control was barely discussed during the World Cup. Quite the contrary, the public seemed to accept it as a necessary part of a much-anticipated event. The widespread public acceptance of exceptional security measures at the football

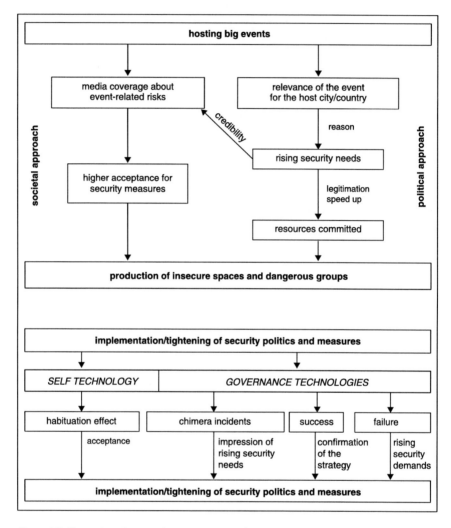

Figure 6.2 Dynamics of event-driven security politics

stadia and public viewing areas, including bag and body searches, and camera surveillance, also contributes to the public acceptance of higher security standards in daily life.

Figure 6.2 schematically represents how large events tend to strengthen security and control politics.

Looking at the right side of the figure, labelled "political approach," one sees that the significance of the event is used to legitimate new security measures, which in turn leads to new resources being committed (e.g. police officers and surveillance cameras). The left side, the "societal approach," shows how media coverage about potential risks and threats operate in the context of the event to publicize, for example, the danger of terrorist attacks or hooligan riots. Such reporting reinforces the need for higher security measures and contributes to the public acceptance of such measures. In sum, such processes produce a form of insecurity that concentrates on certain areas/spaces (like public viewing areas) and is targeted towards certain groups such as Muslims. Demand for security and control increases, which in turn legitimizes intensified security measures. At the individual level, enhanced security measures foster increased self-regulation both during the event and in everyday life. As people adapt to these new, higher-security levels, their self-regulation prepares the ground for new rounds of escalation in security and control instruments.

The use of these new measures is independent of their success or failure, as both success and failure can be used to argue to enhance security standards (Baasch 2009; Haggerty 2009). Excessive security not only leads to a higher detection rate but also to an increase in public complaints. Growing numbers of complaints may, in turn, be interpreted as due to increased levels of disorder due to how those incidents can influence the statistical crime rates (chimera incidents). Consequently, officials can even appeal to peaceful events to argue for more security measures. Similarly, security failures can be attributed to the lack of adequate material and legal resources. In either case, security tends to be enlarged and widened. These dynamics point to the formation of a permanent state of control, which Deleuze (1990) considered symptomatic of control societies.

Conclusion: event-driven security dynamics

While most mega-events occur in cities in the north-west hemisphere of the globe, more cities of the global south are now competing to host these events. The 2010 World Cup in South Africa was the first World Cup in Africa and predictably security was a central concern for this event. And although the framework conditions for Germany and South Africa differ greatly, I anticipate that the South African event will also produce a legacy of intensified security measures. In contrast to the German situation, the main security problem in South Africa was understood to be street crime, and in the lead-up to the World Cup it seemed that quite "traditional" and high-profile strategies were being used to fight this problem. Security resources were topped up to the point that it was reported that: "The South African Police Service will spend R640 million on the deployment of 41,000 officers specifically for the event. Recruitment and event-specific training for this force are under way. The SAPS is on a massive recruitment drive to increase general police numbers by 55,000 to over 190,000 by 2009. The number of police reservists will also double before the World Cup, from 45,000 members to 100,000" (Donaldson and Ferreira 2009: 9).

Secondly, the South African police strategy was expected to be more oriented towards high-profile strategies and zero-tolerance approaches. Critics have claimed that the crime security plan for the tournament was merely a "strong arm" tactic plan following the "iron fist" approach of the existing National Crime Combating Strategy. They also feared that it would negatively affect democratic freedoms and doubted that such a police-state approach could be sustained after 2010 (Horn and Breetzke 2009: 30).

These high-level police strategies also resulted in marginalized groups being evicted or displaced. In her case study of urban planning in the surroundings of the Ellis Park World Cup stadium in Johannesburg, the geographer Bénit-Gbaffou (2009) emphasizes that prior to the event the World Cup was being used to justify strategies that excluded and displaced poor and homeless people. During the World Cup the police took over the security operation at four stadia (Soccer City and Ellis Park in Johannesburg, Green Point Stadium in Cape Town and Moses Mabhida stadium in Durban) after the security company dismissed their guards for striking for higher wages (Basson and Tolsi, 2010). As the South African media reported, the police quelled the protest by using tear gas, rubber bullets and stun grenades, "[. . .] an image all too familiar in township demonstrations over poor delivery of basic services, but which World Cup organisers must have prayed would stay right there" (Smith 2010).

Though the threats may differ, security remains at the top of the mega-event agenda. This is a particular concern given that mega-events have become highly standardized both in terms of organizational procedures and framework conditions. This means that the security measures deployed at one event are apt to be reproduced and built upon for future events. Advocates for mega-events emphasize unrealistic hopes for myriad positive outcomes and seek to avoid disruptions at any cost. Although the specifics may differ, some effects are quite international. Large events are often used to push through displacement strategies or other controversial policies in the field of urban development as well as in the field of security. The need to convey an image of success to the global public produces such pressures that only the highest possible levels of event management and security seem adequate. Unfortunately, these dynamics seem to prevent a more rational analysis or calmer assessment of the need for and effectiveness of event-driven security measures.

References

Baasch, Stefanie. 2009. *Herstellung von Sicherheit und Produktion von Kontrollräumen im Kontext von Großevents: Die Fußballweltmeisterschaft 2006 in Hamburg.* Doctoral dissertation. Hamburg.

Bargel, Marco. 2005. FIFA "Fußball-Weltmeisterschaft 2006™ – Deutsche Wirtschaft steht als Gewinner bereits fest." *Postbank Research* 2/2005.

Basson, Adriaan and Niren Tolsi, 2010. "World Cup security shambles." *Mail & Guardian Online*, June 18, 2010. Available at: http://www.mg.co.za/article/2010-06-18-world-cup-security-shambles (accessed June 21, 2010).

Bénit-Gbaffou, Claire. 2009. "In the shadow of 2010: Democracy and displacement in the Greater Ellis Park Development project," in *Development and Dreams. The Urban*

Legacy of the 2010 Football World Cup, edited by U. Pillay, R. Tomlinson and O. Bass, 200–222. Cape Town: Human Sciences Research Council.
Beste, Hubert. 2000. *Morphologie der Macht. Urbane "Sicherheit" und die Profitorientierung sozialer Kontrolle*. Opladen: Leske und Budrich.
—— 2004. "Policing German Cities in the Early Twenty-First Century." Paper to ASA – 99th Annual Meeting, San Francisco, CA, August 14–17. URL: http://www.fh-landshut. de/uploads/_k/7W/_k7W56EpyI5XbIji78KOLg/ASASF_Beste_Final_Version.pdf (accessed December 5, 2010).
Boyle, Philip and Kevin D. Haggerty. 2009. "Spectacular Security: Mega-Events and the Security Complex." *International Political Sociology* 3: 257–274.
Brenke, Karl and Gert G. Wagner. 2007. "Zum volkswirtschaftlichen Wert der Fussball-Weltmeisterschaft 2006 in Deutschland." *DIW Research Notes*. Available at: www.diw. de/documents/ publikationen/73 /56559/rn19.pdf (accessed August 14, 2009).
Bundesministerium des Innern. 1999. Regierungsgarantien zur FIFA Fußball-Weltmeisterschaft 2006 – Zusammenfassung. In: *Fußball-WM 2006. Abschlussbericht der Bundesregierung, chapter C, annex II*, published by Bundesregierung. Berlin.
Burbank, Matthew J., Gregory D. Andranovich and Charles H. Heying. 2001. *Olympic Dreams. The Impact of Mega-Events on Local Politics*. Boulder, CO: Lynne Rienner.
Deleuze, Gilles. 1990. "Post-scriptum sur les sociétés de contrôle." *L'autre journal*, n° 1, 5/1990.
Donaldson, Ronnie and Sanette Ferreira. 2009. "(Re-)creating urban destination image: Opinions of foreign visitors to South Africa on safety and security?" *Urban Forum* 20: 1–18.
Ehrenberg, Eckehart and Wilfried Kruse. 2000. *Soziale Stadtentwicklung durch große Projekte? EXPOs, Olympische Spiele, Metropolen-Projekte in Europa: Hannover, Sevilla, Barcelona, Berlin. Eine Studie der Sozialforschungsstelle Dortmund*. Münster: Lit.
Foucault, Michel. 1975. *Surveiller et punir. Naissance de la prison*. Paris: Gallimard.
Geese, Stefan, Claudia Zeughardt and Heinz Gerhard. 2006. Die Fußball-Weltmeisterschaft 2006 im Fernsehen. Daten zur Rezeption und Bewertung. *Media Perspektiven 9/2006*: 454–464. Frankfurt am Main. Available at: http://www.media-perspektiven.de/uploads/ tx_mppublications/09-2006_Geese.pdf (accessed December 5, 2010).
Giddens, Anthony. 1990. *The Consequences of Modernity*. Cambridge: Polity.
Graham, Stephen, ed. 2006. *Cities, War, and Terrorism : Towards an Urban Geopolitics*. Malden, MA: Blackwell.
Grant Thornton Strategic Solutions. 2008. "2010: R55bn boost for SA economy." *SouthAfrica.info*. Available at: http://www.southafrica.info/2010/economic-impact.htm (accessed August 14, 2009).
Haggerty, Kevin. 2009. "Methodology as a knife fight: The process, politics and paradoxes of evaluating surveillance." *Critical Criminology: An International Journal* 17 (4): 277–91.
—— 2006. "Tear down the walls: On demolishing the panopticon," in *Theorizing Surveillance: The Panopticon and Beyond*, edited by D. Lyon, 23–45. Cullompton: Willan.
Haggerty, Kevin and Richard V. Ericson. 2000. "The surveillant assemblage." *British Journal of Sociology* 51 (4): 605–622.
Hagn, Florian and Wolfgang Maennig. 2007. "Labour market effects of the 2006 Soccer World Cup in Germany." *Hamburg Contemporary Economic Discussions*. Available at:

http://www.uni-hamburg.de/onTEAM/grafik/1098966615/WP2007-08_FH_WM_ Labour_Market_Effects.pdf (accessed December 5, 2010).

Harvey, David. 2000. *Spaces of Hope*. Edinburgh: Edinburgh University Press.

Häussermann, Hartmut and Walter Siebel. 1993. "Die Politik der Festivalisierung und die Festivalisierung der Politik. Große Ereignisse in der Stadtpolitik." In: *Festivalisierung der Stadtpolitik. Stadtentwicklung durch große Projekte,* edited by H. Häussermann and W. Siebel, 7–31. Opladen: Westdeutscher Verlag.

Horn, André and Gregory Breetzke. 2009. "Informing a crime strategy for the FIFA 2010 World Cup: A case study for the Loftus Versfeld stadium in Tshwane, South Africa." *Urban Forum* 20 (1): 19–32.

Horne, John and Wolfram Manzenreiter. 2006. "An introduction to the sociology of sports mega-events." In: *Sports Mega-Events: Social scientific analyses of a global phenomenon,* edited by J. Horne and W. Manzenreiter, 2–16. Malden: Wiley-Blackwell.

Jäger, Siegfried. 2006. Diskursive Vergegenkunft. Rassismus und Antisemitismus als Effekte von aktuellen und historischen Diskursverschränkungen. In: *Historische Diskursanalysen, Genealogie, Theorie, Anwendungen,* edited by F.X. Eder, 239–252. Wiesbaden: VS Verlag für Sozialwissenschaften.

Lenskyj, Helen Jefferson. 2002. *The Best Olympics Ever? Social Impacts of Sidney 2000.* Albany: State University of New York Press.

Luhmann, Niklas. 1999. *Gesellschaft der Gesellschaft.* F.a.M.: Suhrkamp.

MacLeod, Gordon. 2002. "From urban entrepreneurialism to a 'revanchist city'?" in *Spaces of Neo-liberalism. Urban Restructuring in North America and Western Europe,* edited by N. Brenner and N. Theodore, 254–276. Malden, MA: Blackwell.

Maennig, Wolfgang. 2007. "One year later: A re-appraisal of the economics of the 2006 soccer World Cup." *Hamburg Contemporary Economic Discussions 10/2007.* Available at: www.uni-hamburg.de/onTEAM/grafik/1098966615/WP2007-10.pdf (accessed August 14, 2009).

Matheson, Victor A. 2006. "Mega-Events: The effect of the world's biggest sporting events on local, regional and national economies." *International Association of Sport Economists, Paper No. 06-22.* Available at: http://college.holycross.edu/RePEc/spe/Matheson_MegaEvents.pdf (accessed August 14, 2009).

Naucke, Wolfgang. 2003. *Über die Zerbrechlichkeit des rechtsstaatlichen Strafrechts. Materialien zur neueren Strafrechtsgeschichte.* Baden-Baden: Nomos.

Pilz, Gunter A. 2007. Evaluation und wissenschaftliche Begleitung des Fan- und Besucherbetreuungskonzeptes der FIFA WM 2006. In: *Die Welt ist wieder heimgekehrt. Studien zur Evaluation der FIFA WM 2006,* edited by D. H. Jütting, 73–98. Münster: Waxmann.

Rahmann, Bernd. 1999. Kosten-Nutzen-Analyse der Fußball-Weltmeisterschaft 2006 in Deutschland—Ausgewählte konzeptionelle Aspekte und Ergebnisse. *Professionalisierung im Sportmanagement: Beiträge des 1. Kölner Sportökonomie-Kongresses,* edited by H.-D. Horch, J. Heydel and A. Sierau, 355–373. Aachen: Meyer und Meyer.

Samatas, Minas. 2011. "Surveilling the 2004 Olympics in the aftermath of 9/11: International pressures and domestic implications" in *Security Games: Surveillance and Security at Mega-Events,* edited by Colin Bennett and Kevin D. Haggerty, 55–71. London: Routledge.

—— 2008. "From thought control to traffic control: CCTV politics of expansion and resistance in post-Olympics Greece" in *Surveillance and Governance: Crime Control and Beyond,* edited by M. Deflem, 345–69. Bingley: Emerald.

Schaffrath, Michael. 2006. Das Fernsehen ist als Sieger vom Platz gegangen—Daten, Fakten und Ansichten zum TV-Hype rund um die WM 2006 in *Fröhlicher Patriotismus? Eine WM-Nachlese*, edited by E. Hebeker and P.W. Hildmann, 91–97. München: Hanns-Seidl-Stiftung.

Schindler, Christian and Stefan Steib. 2005. *WM 2006—Fanartikel oder "Kick" für die Börse?* Mainz: Landesbank Rheinland-Pfalz.

Simon, Bart. 2005. "The return of panopticism: Supervision, subjection and the new surveillance." *Surveillance & Society* 3 (1): 1–20.

Simons, Katja 2003. *Politische Steuerung großer Projekte. Berlin Adlershof, Neue Mitte Oberhausen und Euralille im Vergleich.* Opladen: Leske und Budrich.

Smith, David. 2010. "World Cup embraces triumph and disaster." *Mail & Guardian Online*, June 19, 2010. Available at: http://www.mg.co.za/article/2010-06-19-world-cup-embraces-triumph-and-disaster (accessed June 21, 2010).

Smith, Neil. 1996. The *New Urban Frontier: Gentrification and the Revanchist City*. London: Routledge.

Stott, Clifford J. and Otto Adang. 2003. *Policing Football Matches with an International Dimension in the European Union: Understanding and Managing Risk.* Available at: www.liv.ac.uk/Psychology/staff/CStott/Understanding_and_managing_risk.pdf (accessed August 14, 2009).

Stott, Clifford J., Otto Adang, Andrew Livingstone and Martina Schreiber. 2006. "Policing, Crowd Dynamics and Public Order at Euro2004. A report to the Home Office on the project 'A European study of the interaction between police and crowds of foreign nationals considered to pose a risk to public order'." Available at: http://www.liv.ac.uk/Psychology/staff/CStott/Final_Report_2004.pdf (accessed December 5, 2010).

Szymanski, Stefan. 2002. "The economic impact of the World Cup." *World Economics* 3 (1): 169–177.

Tomlinson, Richard, Orli Bass and Udesh Pillay. 2009. "Introduction," in *Development and Dreams. The Urban Legacy of the 2010 Football World Cup*, edited by U. Pillay, R. Tomlinson and O. Bass, 3–17. Cape Town: Human Sciences Research Council.

Veith, Marine. 2010. "Thousands protest against World Cup spending." *Mail & Guardian Online*, June 16. Available at: http://www.mg.co.za/article/2010-06-16-thousands-protest-against-world-cup-spending (accessed June 17, 2010).

Wilson, James Q. and George L. Kelling. 1982. "Broken windows: The police and neighborhood safety." *The Atlantic Monthly*, Vol. 249. Available at: www.theatlantic.com/doc/print/198203/broken-windows (accessed August 14, 2009).

Chapter 7

Commonalities and specificities in mega-event securitization

The example of Euro 2008 in Austria and Switzerland

Francisco R. Klauser

This chapter draws upon empirical insights into security governance at the 2008 European Football Championships in Switzerland and Austria (hereafter "Euro 2008"). My objective is to explore the commonalities and specificities in the securitization of Euro 2008 in the event's two host nations and eight host cities. The paper also looks into the interactions and interdependencies between local, national and transnational stakeholders and their motivations in security governance at sport mega-events.

This research problem covers a large and complex field of investigation. In recent years, the various forms, effects and problems of trans-scalar collaboration in security governance have been acknowledged both from a general perspective (e.g. Power 2007; Amoore and DeGoede 2005) and from the viewpoint of mega-event securitization more specifically (Samatas 2007; Klauser 2008). These studies have provided insights into the normative weight of best practices provided by security professionals moving from country to country, and from event to event (see Boyle 2011). They have shown local stakeholders to be increasingly exposed to globalized networks of expertise that are helping to reproduce previously tested collaborations and templates in security matters. These arguments are summarized in Samatas' study of the external pressures and "policy penetration" (Bennett 1991: 228) in security matters at the 2004 Athens Olympics:

> Athens, protected only by traditional security provisions, was not initially deemed reliable to host the post-9/11 Summer Olympics. Greece had to build an international security alliance and buy the latest security and surveillance technology made in the United States and the European Union to get support and confidence.
>
> (Samatas 2007: 222–3)

There are many good reasons for understanding sport mega-events as highly visible and prestigious projects, whose securitization is firmly embedded in more or less coercive transnational circuits of imitation and standardization. In this approach, however, the role of local agency, motivation and expertise in security governance should not be underplayed, or ignored completely.

My chapter emphasizes precisely this issue. In what follows, I am interested in differentiating the understanding of mega-event securitization as an externally driven, high-profile operation. Based on empirical insights into the securitization of Euro 2008 in Switzerland and Austria, I advance a number of preliminary arguments regarding the degree of autonomy and maneuverability of national and local stakeholders in the field of mega-event securitization.

The chapter is divided into three main parts. The first part briefly outlines the methodological approach, particularities and special appeal of my case study. The second part investigates five main developments in contemporary security governance at sport mega-events: the technologicalization, militarization, commercialization, globalization and standardization of security/surveillance. Specific examples are cited to illustrate how these developments were reflected in the security "dispositif" (Foucault 2009: 11) for Euro 2008. On this basis, I seek to highlight some of the general forces and processes contributing to the recurrent commonalities in the field of mega-event securitization. At the same time, these examples also point towards the existence of important national and local variations in the securitization of Euro 2008, thus underlining the value of particular national and regional characteristics, agencies and motivations in shaping security governance at mega-events.

The third section adds a local dimension to this discussion. Focusing on the host city of Geneva, I locate the issue of mega-event securitization in the context of a specific range of projects, whose realization was driven by complex interactions and coalitions of interest between local, national and transnational stakeholders. Security governance at Euro 2008 was hence conditioned not only by an array of transnational developments and processes, but also by specific national characteristics and a range of local desires and agencies in security/surveillance matters.

My study draws upon empirical insights provided by a two-year research project relating to the securitization of Euro 2008 in Switzerland and Austria. Facilitated by longstanding research collaborations, my research involved ten in-depth interviews with diverse stakeholders in the securitization of Euro 2008 in Geneva. These stakeholders included the security coordinator of the Euro 2008 stadium in Geneva, security personnel at Geneva International Airport, representatives from the Ministry of Justice in Geneva, the city's security coordinator and police personnel. Furthermore, the research has relied on the extensive study of official documents (minutes of local executive and parliament sittings, executive responses to local, regional and national parliamentary debates, and official documents from police sources and UEFA) and on information gathered from various local, national and international media.

My analysis of Euro 2008 draws predominantly on information collected in Switzerland. Austrian examples relating to the securitization of Euro 2008 are considered only with a view to highlighting specific national differences in the securitization of these events.

Particularities of Euro 2008

The European Football Championships are generally regarded as the third largest recurrent sport mega-event in the world (Stadtpolizei Zürich 2007a). Unlike other mega-events such as the Olympics or G8 summits, Euro 2008 affected not just one urban site, but a network of eight host cities in two countries: Basel, Bern, Geneva and Zurich in Switzerland; and Innsbruck, Klagenfurt, Salzburg and Vienna in Austria. Using Euro 2008 as a case study thus offers ideal conditions to investigate how both local and national approaches to security/surveillance may differ for the same event.

By way of further contextualizing the Swiss case study, two particularly important characteristics of Switzerland are worth emphasizing: Switzerland's position as a non-EU member state on the one hand, and its internal federalism on the other. It is safe to assume that both characteristics not only shaped the organizational structures at Euro 2008, but also facilitated the emergence of specific local and national differences in the event's securitization.

Switzerland's non-EU membership

Staging Euro 2008 in a EU member state (Austria) and a non-EU country (Switzerland) raised important questions with respect to security collaboration between the host nations (Public Authorities Security Sector Coordination UEFA EURO 2008 2007: 50). As early as 2003, the two countries had signed a ministerial declaration on intensified cooperation for Euro 2008. Towards that end, a joint working party on security was established to review various critical issues, ranging from the experiences at earlier sport mega-events to the legal framework for implementing joint measures (Public Authorities Security Sector Coordination UEFA EURO 2008 2007: 49). Despite these efforts, however, security collaboration between Switzerland and Austria was further complicated by the fact that before Euro 2008, Switzerland had not implemented the EU Schengen agreement on European cross-border police cooperation (Schweizerische Depeschen Agentur 2006).

According to several Swiss government reports (e.g. Public Authorities Security Sector Coordination UEFA EURO 2008 2007: 17), the specific challenges arising from Switzerland not being a member of the EU were countered by a range of either pre-existing or specifically crafted arrangements with participating, neighboring or transit countries. For example, a temporary visa agreement was signed between the Schengen nations and Switzerland: the border between Austria and Switzerland remained a border between an EU member state and a non-EU country (thus with border controls in place), but no specific Swiss visa was required for fans from participating Schengen countries. This special arrangement also included a temporary agreement that border controls would be staffed by mixed teams from both countries (Amies 2008).

Thus in many ways, Switzerland's non-membership of the EU resulted in a range of special arrangements and solutions for Euro 2008. However, it should not be forgotten that Swiss–European security collaboration at high-risk events had been tested before on several occasions, such as at the G8 meeting in Evian near Geneva in 2003, and the annual World Economic Forum in Davos.

Federalism

The second Swiss particularity to emphasize goes back to the country's internal federalist structure, which designates a particularly high degree of autonomy to the regional (cantonal) and local (municipal) level. At least two major implications of Swiss federalism are worth mentioning here.

Firstly, Swiss federalism increased both the complexity of internal security coordination for Euro 2008 in general, and the scope for diverging regional/local security solutions more specifically. For example, the novel security technologies for Euro 2008 appealed to the political authorities on no less than three geographic scales: the host cities, the cantons, and the confederation (Fürst 2007: 16).

Secondly, Swiss federalism provided many different channels for critical political engagement with Euro 2008. In the local parliaments of the four Swiss host cities, staging Euro 2008 gave rise to a myriad of political debates. Whilst these debates did not question the viability of Euro 2008 as a whole, they in many ways affected specific local solutions regarding a wide range of event-related issues, including security. I will cite three examples. In Zurich, residents and local shopkeepers, widely supported by conservative parties, challenged (in part successfully) security-related road closures around fan zones. In Bern, Euro 2008 raised more fundamental concerns. There, debates focused mainly on the costs, the controversial militarization of public safety, and the commercialization of public space during the event. The city's budget for Euro 2008 (5.6m Swiss francs) was subjected to a public vote held on 17 June 2007. Although a 52.4% majority of the voters confirmed the city's original budget, the narrow result was seen as supporting the opposition's position. In Winterthur (a non-host city of Euro 2008), on 25 November 2007, 56% of the population voted against the city's plan to stage official public viewing of the event.

On different levels, in different places, on different scales, and for different reasons, Euro 2008 was subjected to a variety of more or less politically institutionalized pressures and debates. Assessing the commonalities and specificities of security governance at Euro 2008 must therefore also acknowledge the direct and indirect effects of these more local, political and popular engagements.

Developments in security governance at sport mega-events

In recent years, security and surveillance at sport mega-events have been subjected to repeated academic scrutiny. Although there continues to be a lack of truly empirical and comparative work in this field, these investigations allow us to understand security governance at sport mega-events as the result of, and the catalyst for, a broad set of developments, ranging from processes of technologicalization, militarization and commercialization to the increasing globalization and standardization of security/surveillance matters. It is worth exploring these developments briefly with a view to the securitization of Euro 2008. My aim here is not only to highlight the existence of important general trends (commonalities) in security/surveillance at mega-events, but also to shed light on some national and regional variations of these trends that were evident at Euro 2008.

The technologicalization of security/surveillance

Recent work underlines the role of sport mega-events as test sites for increasingly complex high-tech surveillance applications (Samatas 2007; Boyle and Haggerty 2009; Yu et al. 2009; Giulianotti and Klauser 2010). The illustrations below confirm the importance of sophisticated novel security technologies at Euro 2008. The deployment of unmanned aerial vehicles (UAV) for crowd control in Swiss host cities is a prime example (Stadtpolizei Zürich 2007b). Since 2005, UAVs have been used in Switzerland to monitor traffic, natural disasters and national borders. Yet the policing of Euro 2008 marked an unprecedented urban-centered application of UAV in the country.

Interestingly, the use of UAVs for urban policing was not allowed in Austria (Österreichischer Rundfunk 2008a). It should also be noted that during Euro 2008, UAVs were employed to monitor only three of the four Swiss host cities. In Geneva, the proximity of Geneva International Airport precluded using drones in the city. Here we see national variations in security governance at Euro 2008, as well as the role played by local (in this case urban) characteristics in shaping event security.

The mobile fingerprint identification system bought by the Swiss police provides a second illustration of high-tech surveillance at Euro 2008. This device was publicized as a logical enhancement to the existing stationary fingerprint identification system used at Swiss border checkpoints (Österreichischer Rundfunk 2008b). In Austria, in contrast, no such technology was used during Euro 2008. This example thus again underlines the autonomy of national decision makers in event security. Furthermore, these differences raise a critical question regarding the measures' proportionality. National differences in the securitization of Euro 2008 were not due to differing risk assessments in the two host nations, but resulted from the countries' socio-political contexts, security governance and the motivations for surveillance.

Other novel security measures at Euro 2008 included RFID-based access control systems for stadia, as well as additional CCTV and new police communication technologies in some host cities (both in Switzerland and Austria). I will return to these examples in the last part of this chapter, with a view to the securitization of Euro 2008 in Geneva.

Commercialization of security/surveillance

Euro 2008 created a range of major business opportunities for companies that provide security-related labor forces, advice and technology. In Switzerland, approximately 3,500 private security personnel (provided by a consortium of private security companies for Euro 2008) spent a total of 25,000 working days to secure team hotels, training grounds, stadiums, public viewing events, and other sites (Projektorganisation Öffentliche Hand UEFA EURO 2008: 49).

Regarding the importance of private expertise in providing, designing, and managing security/surveillance installations at Euro 2008, the following account, taken from my interview with Geneva's coordinator of stadium security for Euro 2008, is typical:

> We are constantly approached by companies trying to sell stewarding or security. Other companies attempt to sell geo-localization, helping to localize agents, etc. There are also companies selling high performance radio-communication systems, headphones . . . everything. Eventually, we are approached by sellers of fences, access systems, etc. [. . .] We've taken time to discuss with some of them [. . .] For example, there was somebody, commissioned by several Israeli companies, offering a whole range of services and technologies. I haven't met this particular person, but I've sent him towards other colleagues. We can't meet everybody; we've got to concentrate on the essential. (Security coordinator, Geneva football stadium, personal interview 14 February 2008; my translation)

This quote reveals the intense marketing and lobbying strategies employed by private companies longing to help securitize sport mega-events. Yet it also sheds light on the authority of local decision makers to decide which people to meet, which information to obtain and which problems to prioritize. The quote thus draws attention to both the standardizing forces inherent in the lobbying activities of private security companies, and the autonomy of local authorities who have some scope to choose the providers and solutions that best meet their needs.

Militarization of security/surveillance

Security/surveillance at mega-events can also be placed within the wider context of "new military urbanism," a term that designates the spread of military techniques and militarized definitions/organizations of urban space, particularly in the context of the War on Terror (Graham 2010).

At Euro 2008, for example, UAVs were supplied by the Swiss Air Forces' industry partner RUAG Aerospace, and operated by the army's Drone Squadron 7 in collaboration with civilian personnel at the military airbase in Emmen (Schweizer Armee 2008). Fifteen thousand military personnel were deployed for Euro 2008 in Switzerland – from the air forces, infantry, military security, sanitary services, and communication services – making the event the biggest military deployment in Switzerland since World War Two (Schweizerische Eidgenossenschaft 2007). In Austria, 3,000 military personnel worked at the event (Österreichisches Bundesheer 2008), which also saw the policing debut of Austria's new *Eurofighter Typhoon*, an air fighter developed by the European Consortium Eurofighter Gmbh in the UK, Italy, Germany and Spain (Hoyle 2008).

It should also be noted that neither Austria nor Switzerland (which are not members of NATO) followed the example of the 2006 FIFA World Cup and the 2004 Athens Olympics (both hosted by NATO member states) in deploying NATO AWAC airplanes for aerial surveillance, which further underscores the relationship between national politics and mega-event securitization.

Globalization of security/surveillance

Sport mega-events provide an exemplary illustration of the globalization of social risks and security threats—such as terrorism, hooliganism and organized crime (Jennings and Lodge 2009)—and of the globalization of security partnerships, norms and agreements. In 1990, Switzerland ratified the "European Convention on Spectator Violence and Misbehavior at Sporting Events" (Council of Europe Convention SR 0.415.3), which promoted the exchange of information and police cooperation for sport events. For Euro 2008, Switzerland had concluded a range of additional *ad hoc* agreements with neighboring, transit and participating countries. Interestingly, several agreements written up for earlier events could be recycled for Euro 2008 (examples include the international treaty for air defense with France, signed for the 2003 G8 summit in Evian near Geneva, and the air defense agreement with Italy, established for the 2004 Turin Winter Olympics). These examples reinforce the point that sport mega-events can be catalysts for more enduring international security collaborations (e.g. Chan 2002).

As was the case during the 2006 FIFA World Cup, the Euro 2008 host cities ensured that foreign fan groups were accompanied by police and fan monitors from their own countries. In addition, Switzerland asked France and Germany to supply up to 1,000 police officers (a total of 5,250 working days) to increase its police contingent for Euro 2008 (Projektorganisation Öffentliche Hand UEFA EURO 2008: 50).

Finally, it is important to acknowledge the strong collaboration between public security authorities and Euro 2008 SA (the official organizer of the event, as a subsidiary company of UEFA). Indeed, security in stadia and official UEFA fan zones was delegated to Euro 2008 SA, which accomplished this task by contracting with private security companies (Public Authorities Security Sector Coordination 2007: 47). In studying the commonalities and specificities in security governance at sport mega-events, therefore, it is necessary also to consider the important role played by the event organizers (UEFA, FIFA, IOC, etc.).

Standardization of security/surveillance

The previous sections suggest that security governance at Euro 2008 resulted from complex coalitions of authority and expertise linking together multiple public and private, local, national and international security players. While acknowledging these complexities and interactions, I wish to now stress the weight of globally calibrated best practices in security governance at mega-events.

Perhaps the most obvious example of such standardization relates to the organization of so-called public viewing events, i.e. specifically designed and enclosed "fan zones" where supporters of different nations could drink and party whilst watching the matches on giant television screens. Closely monitored by CCTV cameras, private security agents and police forces, these pre-defined perimeters concentrated fans and helped regulate social life during the World Cup. The concept had its breakthrough at the 2006 FIFA World Cup in Germany, which marked a step change in the large-scale restructuring of event cities into spatially extended fan zones (Schulke 2006). The organization of public viewing zones has since become a bid requirement for both FIFA World Cups and European Football Championships (UEFA undated).

Thus, hosting mega-events implies accepting a series of more or less coercive norms and mechanisms, which help reproduce previously tested and subsequently standardized templates. Further examples include specific policy handbooks and guidelines (such as the "2004 EU Handbook on securing against terrorist acts at major sporting events"), standardized norms and procedures from the bidding process to the staging of the event, progress monitoring by the organizing bodies, and also a range of more informal mechanisms facilitating "institutional learning" and "fast policy transfer" (Peck and Theodore 2001; Boyle 2011) from event to event (technology fairs, expert conferences, exercises, etc.).

Although I mentioned earlier the commercialization of security/surveillance in sport mega-events, it is important to reiterate the role of private companies in standardizing security governance at sport mega-events. By way of example, consider the following Siemens advertising copy:

> Siemens delivers complete infrastructure solutions for major sport events all over the world. Examples are the Olympic Games 2004 in Athens, the Asian Games 2006, the European Soccer Cup 2004 in Portugal or the Soccer World Cup 2006 in Germany, where Siemens equipped all twelve stadiums with latest technology. In Portugal various Siemens Groups bundled application knowledge and synergies in the field of sport infrastructures and contributed most advanced technologies to nine of the ten stadiums.
> (Siemens 2007)

This quotation is a telling testimony of the role and responsibility of private companies, who travel from place to place and from event to event with pre-established plans and designs (exemplars), thus pushing towards increased imitation and policy convergence.

Local specificities in mega-event securitization: the host city of Geneva

My discussion has so far highlighted some of the general trends (commonalities) in security governance at sport mega-events. These examples also shed light on the various regional and national expressions of these trends. I now turn my attention to the host city of Geneva, in order to accentuate how the securitization of Euro 2008 responded to specific local projects and motivations. My investigation focuses on three examples: stadium security, the renewal of Geneva's radio communication network and the extension of CCTV in Geneva's city center. These examples allow me to further examine the reasons for local variations in the securitization of Euro 2008.

Stadium security

As mentioned previously, stadium security at the European Football Championships was mandated to Euro 2008 SA. Constructional, technical, organizational and operational security aspects in and around the event stadia thus fell under the authority of UEFA (Public Authorities Security Sector Coordination UEFA EURO 2008 2007: 46). In Geneva, for example, UEFA's "Binding Safety and Security Instructions" not only prescribed that novel access control measures be used (including RFID and X-ray technology) and that fallow land surrounding the stadium be reallocated, but also provided detailed instructions for stewarding, house rules, and so on (Viot et al. 2009).

The example of stadium security also testifies to the relationship between UEFA instructions and locally anchored motivations in security matters. The following quote from the security coordinator of the Geneva Football stadium helps to make this clear:

> Thanks to Euro 2008, many things have changed around the stadium. For example, there was this building we wanted to disappear for years, which is now being demolished. [...] Beneath the building, installations for hooligans and delinquents are being built. Suddenly, money was made available to do this. So this will remain afterwards. This is something the police had wanted from the beginning, but never managed to get. Now, with Euro 2008, bingo, it's alright. Myself, I've also submitted a proposal for new radios. Quite a few things on the technical and material level will remain in place. (Security coordinator, Geneva football stadium, personal interview 14 February 2008; my translation)

Considering that UEFA instructions prescribed the installations for hooligans and the reallocation of fallow land surrounding the stadium, thus constituting a common feature in all event stadia, the above quote shows how external security directives fused with longstanding demands from police and stadium personnel. Stadium security during Euro 2008 resulted from a complex set of initiatives that were more or less standardized and coercively guided by UEFA, yet also more or less actively initiated and propelled by local stakeholders. Specific decisions resulted from processes involving a range of actors, guided by converging goals, acting from mutually enhanced positions and driven by shared benefits, but also pursuing their own specific agendas and personal projects. The security coordinator's proposal for new radios, a personal initiative for a specific local security solution, is a particularly telling example.

These interactions and interdependencies not only shaped stadium security during the event, but also had an impact on its long-term ramifications, as we see from the security coordinator's mention that some of the changes made in and around the stadium would remain after Euro 2008. Although this conclusion is worth pondering—other security measures were removed after the event, such as X-ray machines for access control—the quote provides an initial glimpse into the security legacy of Euro 2008. Only by recognizing the interactions between transnational, national and local actors can we understand the long-term effects of security governance at sport mega-events.

Radio communication

Compared to stadium security, efforts to renew Geneva's radio communication network for safety and security organizations (known locally as the project "Polycom") touches on issues of much larger scale and complexity. I shall here examine the relationship between Polycom and Euro 2008.

Polycom is a nationally coordinated and subsidized project that aims to integrate the various cantonal radio systems in Switzerland into one unified communication network. The project intends to link all public safety and security organizations on the national, cantonal and municipal level (Bundesamt für Bevölkerungsschutz 2009: 1–2). The national objective is for all cantons to be operational by 2012. Given the number of parties co-financing the project (including the confederation, the 26 Swiss cantons, the affected municipalities and others), the renewal of each sub-system must also be approved by the relevant cantonal parliament.

In Geneva, policy documents on Polycom (cost estimates, statements from relevant authorities, minutes of commission meetings, etc.) have recently been declassified, thus allowing for a study of the project's relationship with Euro 2008. The Geneva Finance Commission's report, outlining the reasons for adopting Polycom on June 13th 2007, is particularly revealing:

> It should be noted that security [. . .], which will affect the image of our canton in the media worldwide, was present in the mind of the members of the Finance Commission. Security depends on this type of equipment,

allowing the networking of all partner authorities and organizations in security and health. In an attempt to provide equipment to the police in an extremely short period, Bill 9901 (allocating the budget for a feasibility study) was quickly dropped. Instead, Bill 10034 (allocating the budget for the project itself) was presented urgently to the Grand Conseil [Cantonal Parliament] on June 14th. Despite the bill's indication on page 16/32 that project planning probably wouldn't allow the running of Polycom in its totality at Euro 2008, we hope that a financial effort of this type (15,266,715 Swiss francs), agreed in these difficult times, would be more than enough to encourage the Grand Conseil to do everything possible to assure a maximal use of this new system during the football matches. (Grand Conseil de Genève 2007: 1; my translation)

For several reasons, the quote above is hugely suggestive. First, it provides ample evidence of the association of Polycom with Euro 2008. Because of Euro 2008, the project had to be adopted "urgently" and without a feasibility study. "Everything possible" was to be done in an "extremely short period" to assure Polycom's "maximal use" at the event. That Polycom was facilitated and accelerated through Euro 2008 is obvious.

Second, the quote indicates that security governance at mega-events strongly interacts with more or less emotionally and/or rationally driven local motivations and concerns. Although more detailed investigation would be required to show exactly how the urgency and unavoidability of Polycom was discursively constructed, the quote exemplifies that the interrogation of the commonalities and specificities of mega-event securitization must also be oriented around questions relating to the complex socio-political dynamics unfolding locally from the event's catalyzing forces.

Given the national importance of the project, it is almost certain that Polycom would have been adopted even without Euro 2008, but there is good reason to believe that the debates would have been very different and the decisions much slower. The pre-assessment of Polycom in 2006 by the Treasury's Finance Inspectorate raised a series of financial concerns, was critical of Polycom's total dependence on Siemens—the exclusive supplier of the network and of its terminals—and questioned the need to replace Geneva's radio system (renewed in 2000 for 10.6m Swiss francs) after just seven years of service (Inspection Cantonale des Finances 2006). In the context of Euro 2008, such doubts were quickly set aside.

Third, as with the example of stadium security, the quote above also tells us something about the legacy of Euro 2008. As we see, Polycom was never expected to be fully operational at Euro 2008. In fact, Geneva police did not start acquiring Polycom radios until December 2009, one and a half years after the event (Office fédérale de la protection de la population 2010: 2). Thus, in reality, Polycom was nothing *but* a legacy. In principle, however, this "delay" could hardly have come as a surprise, given that it took three years to install Geneva's previous radio system (Inspection Cantonale des Finances 2006) and that transferring the police to Polycom was also combined with a large-scale restructuring of the canton's emergency response and security management system.

Ultimately, Polycom also tells us something about the regional and local specificities in security governance at Euro 2008. During the event, it is interesting to note, Polycom was used by the police in Bern and Basel, but not in Geneva and Zürich. In Austria, an altogether different radio communication system was employed.

Police CCTV

My third example relates to the project known locally as "Cyclops," which provided the police with an additional 33 video surveillance cameras for use in Geneva (worth 4.7m Swiss francs). Again, my discussion will be informed by policy documents relating to the discussions and reviews of the project by the Geneva Parliament and its relevant sub-committees.

Cyclops was to have two main video surveillance applications. Twenty-five of the cameras were expected to monitor diverse diplomat sites across Geneva. The other eight were intended for city center squares "at risk" and explicitly related to Euro 2008.

> The objective is to install Cyclops cameras for Euro 2008, at least in the identified problem areas. If data protection legislation is approved in time for the event, the legal framework will be in place for this operation. If not, the parliament will take provisional decisions until the law is voted definitely. (Grand Conseil de Genève 2008: 4–5; my translation)

This quote provides a good starting point for considering some of the event-related logics underlying the implementation of Cyclops. The example again underlines the catalyzing force of Euro 2008: Despite the absence of relevant (cantonal) data protection legislation—another implication of Swiss federalism, which requires video surveillance by *cantonal* police to be based on an adequate *cantonal* legal basis (Müller and Wyssmann 2005)—Cyclops had to be approved urgently in preparation for the event. As an aside, it is interesting to note that the Geneva parliament only passed this legislation on October 9th 2008, three months after Euro 2008, and that the law did not come into force until January 1st 2010.

In order to further underscore how Euro 2008 facilitated Cyclops, it is revealing to look into the minutes of the Finance Commission's meeting on December 19th 2007, during which the project budget was allocated by 9 to 4 votes (1 abstention). The official statements reveal that four of the nine members of the commission who approved this measure explicitly justified their position with reference to the exceptional needs of Euro 2008 (Grand Conseil de Genève 2008: 6–7). Consider the official party line of Geneva's socialist group:

In principle, the socialist group is not in favor of this project. [. . .] The group agrees with those people that are worried about the surveillance of individuals. However, it is important to accept this bill because of Euro 2008. (Grand Conseil de Genève 2008: 6–7; my translation)

The statement highlights the party's dilemma with respect to Cyclops. It is not clear, however, why its members did not opt for a temporary video surveillance system, an option which was debated briefly in the commission and was adopted in the host city of Basel (Justiz- und Sicherheitsdepartement Basel-Stadt 2008). Cyclops was always presented as a permanent solution in Geneva, a fact that underlines both the security legacy of Euro 2008 and the existence of important local variations in event securitization.

Although policy debates on Cyclops resembled those surrounding Polycom in terms of the event's catalyzing force, a series of important differences must be noted. Firstly, unlike Polycom, Cyclops prompted serious debate in the parliament's Finance Commission, in the parliament itself, and in Geneva more generally. In part, this debate can be explained by the relative skepticism towards CCTV and the marked privacy concerns in Switzerland (Klauser 2006). More importantly, however, the project was harder to justify because it was neither nationally coordinated (as for Polycom) nor imposed by UEFA (as in the case of stadium security). Cyclops was driven primarily by local concerns.

In sum, whilst the previous examples of stadium security and Polycom illustrate respectively how UEFA stipulations fused with local demands and motivations, and how Euro 2008 helped to facilitate the local implementation of a nationally coordinated project that would have been constrained in normal times, Cyclops shows that Euro 2008 also worked as a triggering mechanism to facilitate local security projects that were justified in terms of their local benefits both during and after the event.

Conclusion

This chapter has attempted to move beyond an understanding of local security stakeholders at sport mega-events as passive recipients of externally imposed authoritative orders. In contrast, security governance at Euro 2008 was positioned within a complex field of agencies, driving forces and motivations, including a range of international processes and stipulations, as well as diverse national and local predispositions and impulses in security matters.

On the one hand, this chapter has emphasized the weight of transnational security players, who move from place to place and event to event with pre-defined plans and solutions that push towards increased imitation and policy convergence. It has also examined the impact of best practices, and of instructions from the event's organizing bodies (UEFA, FIFA, IOC, etc.), with a view to explaining some of the recurrent commonalities in mega-event securitization. On the other hand, the paper has provided several examples (from Switzerland's drones and mobile fingerprint identification technology to Austria's Eurofighter Typhoons) that indicate important differences in the securitization of Euro 2008 between its two host nations/eight host cities. These examples also point towards the maneuverability of national/local stakeholders in event securitization.

From the perspective of the host city of Geneva, the chapter has shown how external security instructions fused with local motivations and demands (in stadium security), and how Euro 2008 triggered and facilitated projects and developments driven by various interests and agencies both nationally (Polycom) and locally (CCTV surveillance). The exceptionality of the event further increased the weight and perceived importance of particular security projects, whilst at the same time decreasing the scope for criticism and opposition. The examples hence bear striking testimony to the capacity of mega-events to trigger the financing of new constructional, technological and securitizing projects, a point that is powerfully confirmed by research on other mega-events (e.g. Tomlinson 2009: 100).

Based on these examples, the chapter outlines how socio-political, cultural and other particularities of host nations/cities interact with broader trends and developments in security governance at sport mega-events. It also underscores the need to apprehend mega-event securitization as a combination of processes and projects that bring together various public–private, local, national and transnational actors whose positions are defined by interwoven interests and concerns.

Of course, the specific insights gained from this case study might differ in other host cities of other mega-events. For example, interviewees repeatedly emphasized that local stakeholders in security governance enjoy much greater autonomy at the European Football Championships than at events such as the Olympics (which are marked by heavy American involvement and exposed to heightened international pressures). Case by case, from event to event, interactions and relationships between local, national and international security stakeholders vary significantly in terms of the actors, strategies, interests, instruments, and stipulations. Indeed, a central challenge for future research into mega-event securitization will be to undertake detailed and comparative empirical investigations into how different types of events, in different cultural contexts, both resemble and differ from each other.

Such studies will pave the way to a better understanding of the logics and legacies of mega-event securitization and, in addition, may also lay the basis for a critical engagement with security governance at mega-events in terms of the proportionality of the measures employed. In the case at hand, the existence of diverging approaches and dynamics in security matters within the host nations and cities of Euro 2008 underscores the scope for such debates. Policy learning at mega-events should not focus only on the supposed benefits of the measures employed, but also on the possibilities of and benefits from *not* employing certain security measures and particularly intrusive surveillance technologies.

References

Amies, Nick. 2008. "Euro 2008 Co-Hosts: Clear Championship Hurdles—Mostly." DW-World.De. April 20. URL http://www.dw-world.de/dw/article/0,,3253238,00.html (accessed December 12, 2010).

Amoore, Louise and Marieke De Goede. 2005. "Governance, risk and dataveillance in the war on terror." *Crime, Law & Social Change* 43 (2–3): 149–173.

Bennett, Colin. 1991. "How states utilize foreign evidence." *Journal of Public Policy* 11 (1): 31–54.

Boyle, Philip. 2011. "Knowledge networks: Mega-events and security expertise," in *Security Games* edited by Colin Bennett and Kevin D. Haggerty, 169–184. London: Routledge.

Boyle, Philip and Kevin D. Haggerty. 2009. "Spectacular security: Mega-events and the security complex." *Political Sociology* 3 (3): 257–274.

Bundesamt für Bevölkerungsschutz. 2009. *Polycom—Das Sicherheitsnetz Funk der Schweiz*. Information leaflet. July 29. Bern: Bundesamt für Bevölkerungsschutz.

Chan, Gerald. 2002. "From the 'Olympic formula' to the Beijing Games: Towards greater integration across the Taiwan Strait?" *Cambridge Review of International Affairs* 15 (1): 141–148.

Foucault, Michel. 2009. *Security, Territory, Population. Lectures at the Collège de France 1977–1978*. New York: Palgrave Macmillan.

Fürst, Peter. 2007. "Die IT kompensiert personelle Engpässe." *Computerworld* 43: 16.

Giulianotti, Richard and Francisco Klauser. 2010. "Sport mega-events, security and risk management: Towards an interdisciplinary research agenda." *Journal of Sport and Social Issues* 34 (1): 49–61.

Graham, Stephen. 2010. *Cities Under Siege: The New Military Urbanism*. London: Verso.

Grand Conseil de Genève. 2007. *Rapport de la Commission des finances*. PL 10034-A. Geneva. June 19. URL: http://www.ge.ch/grandconseil/data/texte/PL10034A.pdf (accessed March 25, 2010).

Grand Conseil de Genève. 2008. *Rapport de la Commission des finances*. PL 10027-A. Geneva. January 8. URL: http://www.geneve.ch/grandconseil/data/texte/PL1 0027A.pdf (accessed March 25, 2010).

Hoyle, Craig. 2008. "Flying In Neutral—Austria's Eurofighter Typhoons." *Flight International*. August 4. URL: http://www.flightglobal.com (accessed February 2, 2010).

Inspection Cantonale des Finances. 2006. *Rapport relatif au Projet de remplacement du réseau de radio communication de la police (projet Polycom)*. October 24. Rapport No 06–54. Geneva: Inspection Cantonale des Finances.

Jennings, Will and Martin Lodge. 2009. *Tools of Security Risk Management for the London 2012 Olympic Games and FIFA 2006 World Cup in Germany*. Discussion Paper No. 55, London School of Economics and Political Science, Centre for Analysis of Risk and Regulation.

Justiz- und Sicherheitsdepartement Basel-Stadt. 2008. *Temporäre Videoüberwachung während der EURO 2008*. Press Release. May 26. URL: http://www.jsd.bs.ch/newsdateil?newsid=8699 (accessed March 25, 2010).

Klauser, Francisco. 2006. *Die Videoüberwachung öffentlicher Räume*. Frankfurt: Campus.

——— 2008. "Spatial articulations of surveillance at the FIFA World Cup 2006™ in Germany." In *Technologies of Insecurity* edited by K. Franko Aas, H. Oppen Gundhus and H. Mork Lomell, 61–80. London: Routledge.

Müller, Markus and Ursula Wyssmann. 2005. Polizeiliche Videoüberwachung – Rechtssetzungszuständigkeit nach bernischem Polizeigesetz, *Bernische Verwaltungsrechtsprechung 2005* 12: 529–554.

Office fédérale de la protection de la population. 2010. Le canton de Genève est en réseau. Communication OFPP. February 23. URL: http://www.bevoelkerungsschutz.admin.ch/internet/bs/fr/home/themen/polycom/aktuell.html (accessed March 25, 2010).

Österreichischer Rundfunk. 2008a. Risikofaktor Drohne bei der Fussball-EM. *Futurezone. ORF.at*. May 19.

—— 2008b. Mobile Fingerprint-Erfassung bei der EM. *Futurezone.ORF.at*. June 20. URL: http://futurezone.orf.at/stories/286928/ (accessed February 20, 2010).

Österreichisches Bundesheer. 2008. *EURO 2008: Erfolgsbilanz des Bundesheeres*. Press Release. June 30. URL: http://www.bundesheer.at (accessed March 3, 2010).

Peck, Jamie and Nik Theodore. 2001. "Exporting workfare/importing welfare-to-work: Exploring the politics of third way policy transfer." *Political Geography* 20 (4): 427–60.

Power, Michael. 2007. *Organized Uncertainty: Organizing a World of Risk Management*. Oxford: Oxford University Press.

Projektorganisation Öffentliche Hand UEFA EURO 2008. 2008. *Schlussbericht EURO 2008*. Bern.

Public Authorities Security Sector Coordination UEFA EURO 2008. 2007. *The National Swiss Security Strategy for UEFA EURO 2008*. Bern: Schweizerische Eidgenossenschaft.

Samatas, Minas. 2007. "Security and surveillance in the Athens 2004 Olympics: Some lessons from a troubled story." *International Criminal Justice Review* 17 (3): 220–238.

Schulke, Hans-Jörg. 2006. *Fan und Flaneur: Public Viewing bei der FIFA-Weltmeisterschaft 2006*. URL: http://www.hjschulke.de/documents/public_viewing_muenster.pdf (accessed March 3, 2010).

Schweizer Armee. 2008. *Bulletin Flugplatzkommando Emmen*. No 2. Bern: Schweizerische Eidgenossenschaft.

Schweizerische Depeschen Agentur [sda]. 2006. Oberster Polizist warnt vor Verspätung bei SIS. *NZZ online*. September 18. URL: http://www.nzz.ch/2006/09/18/al/newzzes8p7yr3-12_1.61370.html (accessed February 20, 2010).

Schweizerische Eidgenossenschaft. 2007. *Bundesbeschluss über den Einsatz der Armee im Assistanzdienst zur Unterstützung der zivilen Behörden anlässlich der Fussball-Europameisterschaft 2008* (UEFA EURO 2008). Bern: Bundesversammlung der Schweizerische Eidgenossenschaft.

Siemens. 2007. Integrated Solutions at Alvalade Stadium, Lisbon. Press Release. October 16.

Stadtpolizei Zürich. 2007a. UEFA Euro 2008™: Teilprojekt der Gastgeberstadt Zürich—Stand der Vorbereitungen. Press Release. July 13. URL: http://www.stadt-zuerich.ch/internet/mm/home/mm_07/07_07/070713c.htm (accessed February 20, 2010).

—— 2007b. Polizei testet Einsatz von http://www.polizeinews.ch/ Drohnen für Euro 2008. Press Release. November 27. URL: www.polizeinews.ch/ Polizei+testet+Einsatz+von+Drohnen+fuer+EURO+08/370723/detail.htm (accessed February 20, 2010).

Tomlinson, Richard. 2009. "Anticipating 2011." In *Development and Dreams: The Urban Legacy of the 2010 Football World Cup*, edited by U. Udesh, R. Tomlinson, and O. Bass, Cape Town: HSRC Press.

UEFA. undated. *UEFA European Football Championship Final Tournament 2012. Phase I Bid Requirements*. URL: http://www.uefa.com/newsfiles/279728.pdf (accessed March 3, 2010).

Viot, Pascal, Basile Barbey, Valérie November and Hanja Maksim. 2009. "Gérer la sécurité dans les stades: entre ordre prescrit et logiques situationnelles." *International Review on Sport and Violence* 3 (3): 3–9.

Yu, Ying, Francisco Klauser and Gerald Chan. 2009. "Governing security at the 2008 Beijing Olympics." *International Journal of the History of Sport* 26 (3): 390–405.

Chapter 8

Gran Torino
Social and security implications of the
XX Winter Olympic Games

Chiara Fonio and Giovanni Pisapia

This chapter examines issues relating to the security of the XX Winter Olympic Games, which took place in Turin, Italy, from the 10th to the 26th of February 2006. In doing so it explores three crucial aspects of those Games: first, the general social context; second, forms of resistance directed against the Winter Games; and third, the Integrated Security System (ISS) which was developed for and implemented at the sporting venues. This included installing a number of surveillance technologies that sought to address distinctive security concerns. Our aim is to emphasize not only the trend towards the securitization of mega-events at a national and international level, but to also understand some of the broader social implications of the Olympics. In the process, our work also contributes to the literature on the XX Winter Games which to date has tended to be quite limited and fragmented (Bondonio and Campaniello 2006a; Bertone and Degiorgis 2006).

We have chosen to emphasize these factors because of some of the unique peculiarities of the XX Winter Olympics. One of the most important of these concerns the smaller scale of the Turin Games, which translated into a series of security measures that contrast with the typical overzealous security planning for mega-events. So, this smaller scale meant that the security dynamics, surveillance practices and legacies tended to be more limited than the types of intensive security systems implemented in Athens (Samatas 2008) or Beijing (Boyle and Haggerty 2009: 266). That said, while security in Turin was neither exceptional nor "spectacular" (Boyle and Haggerty 2009), it did result in a process of Olympic-motivated urban regeneration (Hiller 2000) that included implementing new surveillance measures. While this could not be characterized as the construction of "defensive urban landscapes" (Coaffee 2003: 6), it still entailed a series of interesting dynamics and proved to be a catalyst to make the city more competitive at a national and international level. These were also planned and implemented in such a way that they generated social conflict in the local mountain communities who resented not being consulted on such matters (della Porta 2008).

Both local and global dynamics shaped the security apparatus for these Games in important ways. At the global level, the XX Winter Olympics were held only seven months after the 7 July 2005 bombing in London. While security planning for the Olympics was in place before July 2005, the London bombings had a key impact on new national legislation, and more generally on security and risk perception shortly before the Games. The attack influenced the dynamics of national crisis management and security measures and led to the introduction of new anti-terrorists laws.

Security planners also had to deal with a host of local protests and demonstrations. Indeed, unlike other mega-events where fears of international terrorism dominate, the situation in Turin was unique, in part, because of how specific local sociopolitical concerns, more than global threats, dominated the overall approach to security. Specifically, security planners were anxious about the prospect that local protests and demonstrations would disrupt the Games. The existence of such issues prompted Johnson to argue that notwithstanding the comparatively small scale of these Games, organizers faced "a local and international picture of potential threats that was every bit as complex as those that faced their predecessors in Sydney, Salt Lake City and Athens" (Johnson 2008: 6).

Thus, the XX Winter Games reproduce a series of global processes and local dynamics which also reveal local and national attitudes towards a top-down approach to secure or "improve" physical environments that proceeded without input from local citizens. In such situations, where the needs of locals become secondary, tensions are likely to emerge (Degen 2004: 141), and this appears to have been the case in Turin.

The Olympic theater

The Olympic theater was divided into the metropolitan area of Turin and mountain area of Val di Susa and Val Chisone. In recent decades, Turin—the first capital of Italy (1861–1864)—has undergone a major transition from being a large industrial city known for being the home of FIAT to emerge as an extremely vital cultural center. This change of identity, due in part to a de-industrialization process, is well reflected in the city's urban transformation (Bondonio and Campaniello 2006a). Given continuing aspirations of greater urban development, it was no surprise that in 1998 the city chose to bid for the Winter Games. These Games were seen by city officials "as a unique opportunity to schedule and accelerate changes in line with the prospects defined by its Strategic Plan" (Bondonio and Campaniello 2006a: 4). This included a desire to integrate the metropolitan area into the international system, promote the image of the city and improve urban amenities. Many of these improvements had long been planned for but, as is so often the case, the Olympics were recognized as being a powerful catalyst for change, a chance to smooth the aforementioned urban transition and to present the city to the international community.

The political motivations of residents in the mountain areas, in particular that of Val di Susa, which connected the main venues, were of concern for the authorities both before and during the Olympics. The major issue here was the proposed high-speed train (Treno Alta Velocita – TAV) project that was designed to align the operations of the Italian rail network to European rail standards, improve travel times both between Italian cities and some European rail networks and culminate in a high-speed network across the entire country. Mountain communites saw this as a controversial development, and for residents of Val di Susa in particular, this was a longstanding issue, as they had commenced a fight against the construction of a new railway in response to a proposal in the late 1990s to connect Turin to the French city of Lyon. Many opposed these initiatives due to anxietes about their overall environmental impact, including the fact that the high-speed train would pass through mountains where asbestos and other dangerous substances would be released during construction.

Officials were anxious that demonstrations against the TAV would spill over into Olympic celebrations which could lead to potential conflicts between police and demonstrators. This anxiety is apparent from the fact that the Anti-Terrorism section of the Division of General Investigations and Special Operations (Divisione Investigazioni Generali e Operazioni Speciali – Digos) characterized the pre-games protests by civil society groups against sponsors and in opposition to the proposed high-speed train as "noteworthy circumstances." In December 2005, the Italian government delayed the start of construction to calm the environmental protests carried out in Val di Susa and Turin. The situation in the valley was in fact very tense. At that time clashes erupted in the town of Venaus in Val di Susa when police first attempted to clear a site occupied by protesters. This was one of the most significant moments in the conflict between police and mountain communities, with the communites complaining about state aggression, injustices and increased "militarization of the valley" (della Porta 2008: 19).

The operational dynamics of these transporation systems introduced some of the Games' most significant lasting surveillance legacies as the "transportation improvement plan" involved a considerable use of surveillance technologies. The SI Pass, for instance, was a radio frequency identifier (RFID) card that could be used to pay for public transport, among other things. The card, developed by a Norwegian company on behalf of the Italian transport operator SITAF, consisted of a double microchip with a double interface (contact and contact-less) utilizing tag and beacon technology. The SI card could be used to avoid using cash on motorway tolls and also enabled transportation users to electronically pay for public city transport (buses, trams and underground) and for using car parks. During the Winter Olympics, the card was also used to pay for ski-passes and could provide access to various events (European Technology Assessment Group 2007).

According to a study written by the Scientific Technology Options Assessment (STOA European Parliament 2007), the SI Pass, similar to the London Oyster Card, provided users with a certain level of identity management which allowed them to gain some control over the use of personal information. It also raised two main concerns: firstly, the users "[had] to do something for [identity management]," and secondly, "it [was] not clear what information [would] be collected besides data on the movement of vehicles" (2007: 61).

Moreover, in order to monitor the vehicle flow in the urban and the mountain areas, the Traffic Operation Centre (TOC) collected data from several devices including transport software (real-time planning, dynamic assignment, "supervisor" system, etc.), monitoring tools (sensors, indicators, and cameras) and centralized traffic lights system. The overall approach was to extend existing ITS systems to monitor traffic and collect information. The TOC organization was particularly complex and consisted of monitoring, collecting and sharing information among the local police, the road police and the motorway companies through an integrated system for traffic, transport management and control.

Contact-less technologies were also implemented in the Turin Metro (underground) with rigid contact-less card tickets containing embedded microchips for remote scanning. Even though the new underground system was not conceived specifically for the Games, it is likely that Turin would never have had such an advanced system without the funds allocated to the Olympics by the Italian Parliament. The Turin Metro contains 550 cameras with images displayed in real time on screen-control-room walls and stored for up to seven days. Of these 550 cameras, 184 are on board the trains while the remaining ones are in the stations. This video surveillance system had to be ready in time for the Olympics for public safety reasons. In addition, the integrated Automatic Light Vehicle (VAL) system which was used for the first time in Italy in the Metro, is another surveillance legacy of the Winter Olympics. It consists of automated driverless light vehicles fully equipped with surveillance cameras.

The Olympics also provided an opportunity for a 90 million euro renovation of the airport (Study Department of the Turin Chamber of Commerce, 2005). In January 2006, the renovated Turin Caselle Airport was inaugurated complete with improvements that included, amongst other things, a new baggage handling system (BHS) equipped with highly sophisticated x-ray control as well as a new terminal dedicated to charter passengers.

Overall, surveillance and security measures were implemented both to deal with new challenges posed by the Winter Olympic Games and to promote a new image of a city which, as noted above, was "in transition." The aforementioned measures reduced the level of "surveillance free areas" in a city that at that time had considerably less surveillance technologies compared to Milan (Fonio 2005) or Rome. One major issue of concern, however, is that the Olympic legacies related to surveillance went far beyond the Games and received almost no media attention.

"No Olympics!"

The other key local group that was of concern to the authorities was the "Against Olympics Coordination" (now referred to as "No Olympics Committee" or "No Olympics!"). This group was founded in 1997 during the World Alpine Sky Championship held in Sestrière, Italy, when the first rumors emerged that Turin would be applying to the IOC to host the XX Olympic Winter Games. "No Olympics!" was formed from various civil society groups such the Askatasuna Community Center (Turin), organizations for the conservation of the environment (i.e. World Wildlife Fund) and non-governmental organizations (i.e. Amnesty International). Paramount amongst the concerns of such groups was the detrimental environmental and economic impact of the Olympics. These groups saw the new infrastructure built in Val di Susa as a potential risk to an already damaged environment (Bertone and DeGiorgis 2006). They also saw these developments as a waste of public money ("public expense for private gain") particularly given that, according to "No Olympics!," it was unclear how the massive investments in infrastructure would be used in the future. The types of environmental concerns expressed by "No Olympics!" have subsequently been voiced at other Olympic Games, including by "No 2010 Olympics on Stolen Native Land," a resistance network that opposed the 2010 Winter Olympics in Vancouver.

Although it would be misleading to suggest that the "NO TAV" group and the "No Olympics!" committee were the same civil society group, the aims of their protests were, to a certain extent, similar. Residents of Val di Susa and supporters from many places often gathered together to prevent the TAV by occupying construction sites and demonstrating to try and block the Olympics. Moreover, solidarity and support for the mountain communities was part of the struggle against the Winter Games carried out by "No Olympics!" Peaceful demonstrations, occupations and actions built a strong sense of community in the valley and it is likely that they opened areas for other debates, such as those pertaining to the environmental impact of numerous works (sporting venues, road infrastructure and villages for the athletes and press) for the Games.

Olympic sponsors also faced harsh criticism that the Games were not sustainable. Finmeccanica was a particular target of such concerns. Finmeccaica is a major Italian industrial group operating in the aerospace, defense and security sectors and Elsag Datamat, one of its subsidiaries, designed and implemented the integrated security system at the Winter Games. The "No Olympics!" group claimed that the role played by Finmeccanica went against the ethical principles expressed by the TOROC in the Charter of Intents and broke the IOC prohibition on creating sponsorship relations with companies that produce weapons. Moreover, anti-Olympics protesters criticized the increased militarization of the event, which was apparent in the involvement of other official suppliers such as Garret Metal Detectors that, according to the committee, shipped several thousand security devices to Iraq.

As noted, part of the Italian protests surrounding the Olympics pertained to the costs of the event. The financial model chosen for Turin involved private funding for organizational expenses (incurred by TOROC) and a series of public investments (Bondiello and Campaniello 2006b). As the private funding was very limited, most of the funds ultimately came from the Italian government which was primarily responsible for financing the investments including constructing venues, infrastructure, and roadworks. In their Sustainability Report (2006), TOROC estimated the overall games budget updated to October 2006 was 1,239 million Euros. Giulianotti and Klauser claim that the overall cost was US$1.4 billion (2010: 2). Bondonio and Campaniello (2006b) noted that these expenditures made

Turin 2006 an expensive event when compared to the previous eight stagings of the Winter Olympics, as it was "more than 11 times more expensive than Lake Placid 1980 and more than 18 times more expensive than Sarajevo 1984" (2006: 11).

The major expenditure for the Games is officially identified as "technologies" (20%), although it is not apparent what exactly "technologies" means (x-ray machines, surveillance cameras and so forth?). This makes it challenging to estimate the amount of the budget dedicated to security and surveillance devices. One of the security officials we spoke with argued that the Organizing Committee spent around 20–25 million Euros on security technologies (surveillance cameras, control rooms, x-ray machines) at the venues. However, this cost does not include the surveillance system introduced to the public transportation system.

As explained, awarding the XX Winter Olympic Games to Turin was controversial in some circles, especially amongst the extreme left and established urban environmental movements. To promote their positions, those groups organized anti-Olympic protests and also illegally occupied public and private buildings which they used as base camps to coordinate activities designed to disrupt Olympic-related events. One high-profile target was the Olympic Torch relay through Italy, which these two groups used as an opportunity to publicize their grievances. Both the torch-bearers and the sponsors logistical vehicles were targeted during the relay, which included assaults on officials and damage to vehicles. These incidents did, however, take place outside the Olympic venues and were efficiently contained by the Italian police forces. Nonetheless, they were widely reported in both the local and international media, creating a degree of publicity that made some sections of the Italian population uncomfortable about how they would be welcomed at the Games.

The protestors' actions seem to have alienated broad segments of the Italian citizenry. The majority of the population and all the main political and sport-related leaders condemned their behaviour. These two groups ultimately abandoned any intention to protest during the Olympic event in the light of the public criticism, and also due to the increased security measures around the Olympic venues.

The Integrated Security System (ISS)

In advance of the Games, extensive security measures were established to deal with potential security threats. This included a new anti-terror law approved in August 2005 and the Olympic Decree of February 2006 which brought into force "urgent measures to guarantee security during the Winter Olympic Games." As noted, this decree was conceived within the tense post-7/7 political climate when the terror alert level in Italy rose significantly. The "urgent measures" focused specifically on recruiting and hiring law enforcement personnel ("up to 1,115 police officers" and "50 fire units"), along with allocating special funding to develop anti-terrorist security in Piedmont's airports. In December 2005, the Minister of Interior, Giuseppe Pisanu, addressed Parliament on new updated security measures. During his speech, he assured Parliament that law enforcement was developing a broad plan to address public security that consisted of a main Olympic security room linked to a national information center that actively collaborated with intelligence services in various countries and 9,000 police officers. Ultimately, the venues and Olympic villages were staffed by a total of 14,184 personnel (State Police, Carabinieri, Financial Police and State Forestry Corps) who deployed 465 metal detectors, 203 x-ray machines, 948 hand-held metal detectors and 583 surveillance cameras.

Prior to the Games, officials ran through a series of security mock-ups. This included three major counterterrorism exercises which took place in Milan, in Rome (September 2005) and in Turin (October 2005). One of the main factors they were testing was the operation of the "security ring" system, which became a key component in security operations during the Games. This approach divided each Olympic venue into various security areas, each of which had distinctive security characteristics and different access rules. A "venue" was a structure officially identified as one that would stage an activity of the XX Olympic Winter Games and that had to be managed and organized by TOROC. The Olympic venues included both permanent and temporary structures. The former were construction projects developed by various local governments as legacy projects of the Olympic Games, while the latter were set up exclusively for the event as set out by the requirements of the International Olympic Committee (IOC) for each Olympic-related activity.

The most external layer of security at an Olympic venue was referred to as the "Controlled Area," and it consisted of a buffer zone between the actual venue and the surrounding territory. In this area Italian law enforcement personnel conducted random security checks on people, vehicles and goods. It was situated in the public domain and included major roads, commercial activities and buildings. As such, it did not limit access to people, goods and vehicles and was not delimited by a defined perimeter (e.g. fence).

The second security layer around the Olympic venues was the "Soft Ring," characterized by a perimeter that consisted of different road blocks called "vehicle permit checkpoints" (VPCs). This perimeter demarcated a zone that was accessible only to authorized vehicles which had to display a specific pass—a Vehicle Access and/or Parking Permit (VAPP)—affixed to their windshield.

The most important component of the general security system developed for the XX Olympic Winter Games' venues, however, was the creation of the "clean" "Hard Ring" area. This Hard Ring was the third and ultimate security cordon immediately surrounding the Olympic venues. It was physically delimited by a protective barrier and was, from a safety and security point of view, a clean/sterile space where all goods, people and vehicles were security screened before entering by Italian law enforcement personnel. Moreover, there was an additional security area (the "Security Ring") created specifically for the Olympic Villages of Turin and Sestrière. This security ring perimeter was characterized by a high fence and was located inside the above-mentioned Hard Ring perimeter

The Hard Ring sought to keep the enclosed area clean (absent from prohibited or hazardous items) by security screening all goods, people and vehicles entering it. Here, the term "clean" refers to the status of a venue, but it could also be applied to a facility, person, vehicle, goods or material package which was known to be free from prohibited and hazardous devices. Prohibited items were those objects, defined by Italian law enforcement agency as being "intrinsically dangerous" and therefore prohibited within every Olympic venue's perimeter. The prohibited items list included firearms, ammunition, explosives, chemical or incendiary devices and instruments commonly defined as weapons.

Keeping the Hard Ring area clean was accomplished by applying the Olympic venues' access rules and involved two different operations: the venue "Lock Down" and the security sweep conducted by law enforcement. "Locking down" a venue was the first action to be taken in order to create a clean Hard Ring area. It was defined as being a state of security readiness and involved activating all the Integrated Security System (ISS) technologies, which we detail below.

Venue security sweeps were also designed to ensure that each Olympic venue was free from prohibited and hazardous items. They were conducted by Italian law enforcement personnel utilizing special equipment (e.g. explosive detection dogs), in cooperation with TOROC staff, in all the Olympic venues in accordance with the schedule defined in the Olympic Venue Security Sweep Schedule (OVSSS).

The Vehicle Screening Areas (VSAs) were positioned at the perimeter of the Hard Ring area where the security screening of vehicles took place. Those areas were at least 100 meters from the Olympic venues' critical areas. At the VSAs the TOROC staff ensured that each vehicle entering the venue's Hard Ring area had a proper VAPP and that the occupants of the vehicle had a valid accreditation. Italian law enforcement personnel ensured that each vehicle entering the venue's Hard Ring area and its occupants were free from prohibited or hazardous items through a detailed security screen. The vehicles' occupants were also screened through magnetic x-ray and physical bag searches ("Mag and Bag") situated at the VSAs.

Three levels of vehicle security screening were conducted at the VSAs and ranged from level 1 (full screening of vehicles) to level 3 (external screening). Those levels were determined by Italian law enforcement personnel who assessed the information related to each vehicle. This included considerations pertaining to its origin (for example if the vehicle was arriving from another Olympic venue), if it deviated from the designated route, if it was escorted or if it was heading for the venues' Hard Ring area. Again, such screening aimed to detect and deter the introduction of any prohibited or hazardous items (e.g. Improvised Explosive Devices) into a venue's Hard Ring.

"Mag and Bag" security screening was also situated on the perimeter of the venues' Hard Ring area, where individuals could gain entry to the Olympic venues. To enter, spectators had to possess a valid ticket and staff had to have a valid accreditation. In addition, staff and spectators had to be security-screened by Italian law enforcement personnel. The security inspection consisted of screening people and their belongings (bags or hand-carried items). This screening was done through a combination of technological (i.e. walk through magnetic metal detectors) and physical screening (tables for manual inspection of people's bags). Moreover, different Olympic client groups (e.g. athletes, media, spectators, dignitaries) had to use specified "Mag and Bag" points to enter the venues. The same security screening was conducted for each person entering the venue. The only peopled who were not searched in this way were international dignitaries, providing they were escorted by Italian law enforcement personnel.

To implement the security policies and the procedures related to the Olympic venues' access rules for people, vehicles and goods/ materials, TOROC, in cooperation with Italian law enforcement agencies, developed and implemented the Olympic Venues' Integrated Security System (ISS), a combination of various security technologies and physical equipment at the venues' Hard Ring perimeter.

The Integrated Security System was designed to assist security personnel (e.g. Italian law enforcement and TOROC personnel) in implementing the Olympic venues' "Hard Ring clean area" principle. The technologies/physical equipment utilized through the Olympic venues' Integrated Security System (ISS), were put into practice with the Lock Down of the venues and consisted of the following:

- A 2.7 meter high security fence surrounding each venue in the Hard Ring.
- Illumination of the perimeter of each venue in the Hard Ring and additional critical areas (e.g. generators). Such illumination allowed the surveillance cameras and the anti-intrusion detection systems to function and patrollers to carry out their duties.
- Surveillance cameras—A surveillance system that transmitted digital images from fixed and domed cameras to the Venue Security Control Room (VSCR) located inside each Olympic venue. Here a TOROC operator monitored the images of the cameras. The fixed cameras were set up every 60 meters alongside the perimeter of the venues' Hard Ring while the dome cameras were set up at strategic points (e.g. in the "Mags and Bags" areas).
- An Anti-Intrusion Detection System designed to detect any unauthorized entry into the Olympic venues' Hard Ring from possible adversaries (e.g. a terrorist group). This system was managed by a trained TOROC operator in the Venue Security Control Room.
- A Venue Security Control Room (VSCR) which was the center for the coordination and management of TOROC security operations in each Olympic venue. Located there was all of the monitoring equipment related to the surveillance cameras and the anti-intrusion detection systems. The Control Room was run by a senior TOROC staff member who managed the system through two-display videos. The left monitor showed the cartographic map of

the venue and displayed all relevant graphic and textual information, such as the presence of surveillance cameras. The right video allowed the operator to visualize the camera's video flows or to zoom in on a particular situation or object.
- Security patrolling of the Olympic venues' Hard Ring perimeter and critical areas. These patrols were conducted by Italian law enforcement personnel and TOROC staff. The critical areas which received particular attention consisted of all the operational areas (e.g. generators), the spectator seating areas/stands and the field of play in the competitive venues.

Olympic venues' security personnel and the CNIO

The three components of the general security system for the Olympic venues consisted of: (1) policies, procedures and contingency plans, (2) physical and technological equipment, and (3) personnel. Concerning the latter, a total of 9,278 people from Italian Law Enforcement Agencies worked at the Games. Before, during and after the Olympics, each of those individuals had to carry out specific duties informed by the documents that set-out the roles and responsibilities for each organization. TOROC, for example, among other responsibilities, was charged with implementing the security infrastructure of the Olympic venues, securing communication and transport systems, and managing ticket sales and marketing. The central and local law enforcement authorities of the Ministry of Interior were responsible for the event's public safety, and for protecting the Olympic sites, villages venues, athletes, delegates and dignitaries.

The criteria adopted to work out a security plan drew on best practices and lessons learned from past events carried out both in Italy (i.e. Genoa 2008, Naples Global forum 2001) and in other countries, in particular in Japan/Korea (2001), Salt Lake City (2002), Greece (2004) and the 2004 UEFA Championships in Portugal. On the basis of such after-action reports the Ministry of Interior set up the CNIO (National Information Centre on Turin Winter Olympics) that had two main goals: to intensify intelligence activity and improve international police cooperation. In particular the tasks of the CNIO before, during and after the event were, *inter alia*, the following:

- *Before the event*: collect, analyze and exchange information at a national and international level and transmit information to the provincial Public Security Authorities; collect useful information involving all stakeholders (from the media to the sponsors).
- *During the event*: manage information flow to assess potential threats.
- *After the event*: produce a final report focused on security activities.

It is important to draw attention to the security activities that pertained to international police cooperation. If, at a national level, the flow of information involved several security departments (such as border police, intelligence services, etc.), at the international level the information flow involved a much wider network of intelligence services of participating countries, including Interpol, Europol and PWGOT (Police Working Group on Terrorism) and the CASA (Antiterrorism Strategic Analysis Committee). As an example, the permanent contact point of the CNIO had to exchange information focused on risk analysis of potentially threatening people and groups expected to attend the Olympics. The technical surveillance used for gathering and processing intelligence included both the Olympic Control Room Tetra (Radio Communication System) and GIS (Geospatial Information System). Moreover, the Integraph Technology provided by the Italian Military Institute integrated feature data (collected, validated and integrated mapping data taken from multiple sources) to aid in the security operations during the Olympics.

All of these security measures appear to have successfully limited the opportunity for untoward eventualities. The XX Winter Olympic Games took place without any relevant security incident and the security system implemented during the event by the various players (both law enforcement agencies and TOROC) was deemed to have been efficient and effective in delivering an incident-free event. Outside the Olympic venues and in the Olympic territory no incident related to the event took place. Only two bomb threats were recorded by the Italian police in the Olympic territory. These occurred outside the Olympic venues and were assessed as being nuisance alarms. Information retrieved from the Italian police (Questura di Torino) and from TOROC, suggests that no significant incident took place during Turin 2006 at the Olympic venues and in the broader Olympic territory. The registered security incidents that did occur inside the Olympic venues were confined to a group of protesters who tried to gain visibility during some Olympic competitions in order to unlawfully advertise a virtual casino called "Golden Palace." They were unsuccessful and were arrested and questioned by the Italian police. Subsequently, an individual wearing a "Golden Palace" T-shirt, attempted to interrupt the speech of TOROC President Valentino Castellani during the closing ceremony at the Olympic Stadium. The individual was immediately blocked by the police and no significant disruption resulted. Perhaps the most notable criminal development was the theft of some ICT equipment from specific venues. These thefts occurred after the competitions and therefore when the venues were closed to the public. In the majority of cases the people responsible were arrested by the police and the equipment was retrieved.

Conclusions

The national literature focused on the XX Winter Games held in Turin in 2006 event is extremely limited. Hence, our aim here was to illustrate several neglected aspects of those Games. We have described both the integrated security system and other significant issues relevant to surveillance practices of the Olympics. Additionally, this chapter has illustrated how the globalization of threats and international terrorism played a role in national legislation, perceptions of security and risks before the Games, and more importantly, in the development of security measures and crisis management during the Olympics. We also accentuated the social context of the Olympics and resistance in the mountain regions to the Games, which, Della Porta claims, was particularly "eventful" due to the "symbolic and physical struggles around the occupied sites in Val di Susa" (2008: 23).

In terms of the surveillance dynamics that were at play, the Winter Games resulted in surveillance measures at the venues, but also at the urban level, mainly because of efforts to monitor vehicle flows and to improve the transportation system. Significantly, the Turin Metro system was the first in Italy to use automated driverless light vehicles fully equipped with real-time video surveillance. Therefore, the Olympics did increase surveillance far beyond the venues, and these initiatives were portrayed as being necessary in order to renovate the city and to present it to the international community.

It is interesting to note, however, that the main public issue surrounding the Games was that of environmental impact, not surveillance. The installation of surveillance cameras both at the venues and within the urban context, in particular on public transport, did not raise the type of public concerns as occurred in Greece (Samatas 2008). The increased use of surveillance cameras in the main Italian cities, coupled with the lack of public debate around the potential erosion of privacy, is intriguing. That said, our understanding of this development is unfortunately limited because empirical research focused on surveillance in Italy is extremely sparse (Fonio 2005; Calenda and Fonio 2010) in comparison to other European countries.

When placed in a larger political and historical context, it is particularly surprising that Italy, a post-authoritarian society with a history of having its citizens exposed to a massive coercive surveillance apparatus, has almost erased such concerns from the collective memory. Such cultural forgetting arguably shaped the minimal level of resistance against the increased use of monitoring devices. It appears that, as Italy's fascist history recedes from living memory, Italian civil society groups and social movements have redefined their priorities in light of the unique appeal and challenges of events such as the Olympics.

References

Bertone, Stefano and Luca DeGiorgis. 2006. *Il libro nero delle Olimpiadi di Turin 2006*. Genova: Fratelli Frilli.

Bondonio, Piervincenzo and Nadia Campaniello. 2006a. *Turin 2006: An organizational and economic overview*. Working paper WP1/2006. URL: http://www.omero.unito.it/?Working_papers (accessed March 25, 2010).

Bondonio, Piervincenzo and Nadia Campaniello. 2006b. *Turin 2006: What kind of winter Olympic Games were they?* Working paper WP2/2006. URL: http://www.omero.unito.it/?Working_papers (accessed March 25, 2010).

Boyle, Philip and Kevin Haggerty. 2009. Spectacular Security: Mega-Events and the Security Complex. *International Political Sociology* 3(3): 257–74.

Calenda, Davide and Chiara Fonio, eds. 2010. *Sorveglianza e società*. Acireale: Bonanno.

Coaffee, Jon. 2003. *Terrorism, Risk and the City*. Aldershot: Ashgate.

Degen, Monica. 2004. Barcelona's games: the Olympics, urban design and global tourism. In *Tourism Mobilites*, eds M. Sheller and J. Urry. New York: Routledge.

della Porta, Donatella. 2008. *Eventful Protests, Global Conflicts*. URL: http://www.bc.edu/schools/cas/sociology/meta-elements/pdf/EventfulProtest.pdf (accessed February 5, 2010).

European Technology Assessment Group. 2007. *RFID and Identity Management in Everyday life. A study by the STOA, European Parliament*. URL:http://www.itas.fzk.de/eng/etag/document/2007/etag07a.pdf (accessed September 10, 2009).

Fonio, Chiara. 2005. *La videosorveglianza. Uno sguardo senza volto*. Milan: FrancoAngeli.

Giulianotti, Richard and Francisco Klauser. 2010. Security governance and sport mega events: Toward an interdisciplinary research agenda. *Journal of Sport & Social Issues* 34(1): 49–61.

Hiller, Harry, H. 2000. Mega-events, urban boosterism and growth strategies: an analysis of the objectives and legitimations of the Cape Town 2004 Olympic bid. *International Journal of Urban and Regional Research* 24(2): 439–58.

Johnson, Chris D. 2008. *On the Convergence of Physical and Digital Security for Public Safety at Olympic Events*. URL: http://www.dcs.gla.ac.uk/~johnson/papers/Olympic_Security_2008/Johnson_Olympic_Security_2008.pdf (accessed February 5, 2010).

Samatas, Minas. 2008. From thought control to traffic control: CCTV politics of expansion and resistance in post-Olympics Greece. In *Surveillance and Governance: Crime Control and Beyond (Sociology of Crime, Law and Deviance Vol. 10)* ed. M. Deflem, Bingley: Emerald 345–69.

Study Department of the Turin Chamber of Commerce. 2005. *Turin Economy. Report on Turin Province*. URL: http://images.to.camcom.it/f/Estero/To/Turin_Economy_2005.pdf (accessed March 7, 2010).

Sustainability Report. 2006. URL: http://www.unep.org/pdf/TorinoReport.pdf (accessed March 25, 2010).

Chapter 9

Mega-events and mega-profits
Unravelling the Vancouver 2010 security–development nexus

Adam Molnar and Laureen Snider

Academic and political commentators have commonly sought to understand the Olympics as a cultural dynamic, a "spectacle" that motivates certain actors to project their relative interests in localized spaces and as well on a global scale (Hiller 2006; Boyle and Haggerty 2009b). Mega-events, as this argument goes, are monumental cultural events (Roche 2000) that rely on the audacity of spectacle to dramatize and condition the cultural, political, legal and economic landscape. Extending these insights into surveillance studies, Boyle and Haggerty (2009b: 259–260) position spectacle and the disciplinary mechanisms of anxieties associated with mega-events to explain the risk management practices of security planners. The dynamic social implications of the spectacle condition dramatic regimes of securitization and surveillance such that sovereign power emanates from the production and consumption of spectacle. In similar fashion Vida Bajc (2007: 1648) writes that security meta-rituals "demonstrate[s] that the process of transformation of [the] public space [of mega-events] from one of routine of daily life into a sterile area [that] has a ritual form [that] . . . . separates insiders from outsiders and brings about a new socio-political reality." Put another way, the "security-meta ritual" legitimates security and surveillance practices by normalizing the social hierarchies it imposes. Bajc focuses on the over-determination of dividing practices in mega-event security, but the signifying practices associated with capital are absent (perhaps due to her empirical focus on presidential addresses). Klauser (2008: 181) links commercialization and mechanisms of surveillance, but only by foregrounding the significance of "neutralized space" created by granting absolute commercial rights to event sponsors. Neoliberal privatization and its articulation with security and surveillance, however, cannot be reduced to control over sponsorship rights and consumptive practices in particular urban "zones," nor can it be limited by the methodological temporality of the event itself.

We suggest instead that mega-events promote unique *models of development* (Aglietta 1979), often signalling significant shifts in urban development projects, where the investment of billions of dollars in sporting, tourist, transportation, and security infrastructures is privileged over social welfare (Hiller 2006; Hall 2006). These temporary instances of "accelerated development" (Gaffney 2010) produce social transformations favouring the priorities of capital, temporarily and permanently transforming socio-spatial relations, rendering visible stark inequalities in localized contexts, and leaving legacies long after the event has ended (Gaffney 2010; Hiller 2006; Newton 2009; Lenskyj 2008). Mega-events create possibilities for a series of political exception(s), amassing political will and economic backing for major development projects that would not, under other conditions, garner public support.

An over-focus on the Olympics as cultural "spectacle" (Boyle and Haggerty 2009b), or "public ritual" (Bajc 2007) downplays the structural and strategic coupling of *modes of accumulation* with legal frameworks and security-oriented mechanisms of governance—a confluence we call the "mega-event security–development nexus" (MESDN). We argue instead that key political economic/structural contradictions of accelerated development, particularly the impacts of uneven development and disparities in large-scale public expenditures, are exacerbated by mega-events. Mega-events, then, uphold the status quo of a neoliberal and *neoliberalizing* political economic order (Peck and Tickell 2002), intensifying and revealing the displacement of insecurity (such as social inequality and environmental externalities) onto marginalized populations. Simultaneously, they deepen political conflicts over public spending on mega-event developments, developments that are now legitimated by the "necessity" to create the "entrepreneurial city" required for a "successful" mega-event (Harvey 1989). As we shall demonstrate, this is particularly evident in the mega-projects that bolster real estate and tourism; creating in the process a significant imperative for legal restructuring and security spending to establish control over the spaces and sites of the mega-event.

Defining the mega-event security–development nexus (MESDN)

A nexus is defined as "a network of connections between disparate ideas, processes, or objects" that implies an "infinite number of possible linkages and relations" (Stern and Ojendal 2010: 11). This theoretical approach is inspired by Karl Polanyi's theory of economic history (*The Great Transformation* ([1944] 2001) underpinned with a critical realist perspective of the Regulation Approach (RA) (Peck and Tickell 2002; Jessop 1995; 2002) and "post-development" thinkers (Escobar 1995; Pieterse 2000). RA highlights structures and strategies of accumulation and governance, while post-development theory emphasizes social and political inequalities that emerge through neo-liberal models of modernization and capital accumulation.

Security governance is conceptualized here as a technique of governmentality and social control based on inter-organizational collaborative networks (Bigo 2005) operating through socio-political co-ordination. Transcending the traditional public–private divide (Krahmann 2003; Brighenti 2008), it is enacted through " 'tangled hierarchies,' parallel power networks or other forms of complex interdependence across different tiers of government" (cf. Jessop 1995: 310). While the discursive risk management strategies of security organizations may not always explicitly support regimes of accumulation, the technologies of exclusion enacted structurally underscore the juridical order—both in terms of property relations and the commodity form. This raises difficult questions about what particularities of security governance one might consider as a public or private good (Zedner 2003).

In *The Great Transformation*, Polanyi noted how the expansion and deepening of economic markets was simultaneously accompanied by extra-economic forces, involving political intervention—through both state and civil society countermovements—to establish social control and steer the economy "in defence of society" (Hettne 2010). Polanyi's "great transformation" of market society occurs, then, through a "double movement"—the first movement involves capitalist expansion as an intensification of the search for new markets. As social and environmental externalities created by these market-centric forms of (neo)liberalism become obvious, capitalist crisis-tendencies are produced, stimulating a "second movement" to counter the destructive economic consequences from the first movement. The second movement acts through organized political intervention that seeks to resolve instabilities and disorder and regulate the aggressive strategies of capital accumulation. The aim is to protect market society and the natural environment from resistance and from the destructiveness of market capitalism.

The Parisian *régulation* school (RA) provides a methodology to analyse the interconnections between the institutions and dynamics of mega-event-specific regimes of accumulation and security governance. RA highlights the conflictual nature of capitalism, showing how, at particular historical moments, certain economic, political, and social institutions come together and provide the conditions—the "norms, institutions, ... social networks and patterns of conduct" (Jessop 2002: 93) that regulate class conflict and competition. Different accumulation regimes incorporate (and exclude) differing sets of social institutions.

In the context of the MESDN, a strategic and structural coupling emerges between regimes of accumulation (Aglietta 1979) and modes of governance (cf. Jessop 1995). We argue, first, that mega-events provide a site where capital accumulation strategies, legal systems and police-security coincide in time and space, a setting where the profit maximizing strategies of capital and the political regulative goals of the nation state and its key justice and policing institutions reinforce each other. The mega-event, then, gives a range of powerful actors and institutions a momentous opportunity to maximize their institutional goals and regulate (nullify or neutralize) societal resistance (Hiller 2006). This paves the way for capital to pursue infrastructural projects and move toward the "entrepreneurial city" (Harvey 1989). In the MESDN these commercial priorities intersect with legal and security/surveillance practices aimed at regulating the poor, policing dissent, and protecting commercial rights. Thus our second argument is that the mega-event maximizes growth opportunities for the private security industry, transforming the public sector and the security landscape far beyond the fleeting event itself. The imperatives of capital, operating through the private security industry and the perceived need to exponentially expand the homeland-security-industrial-complex, are accompanied by deep transformations in public safety and security governance into multi-tiered, complex inter-organizational hierarchies. This chapter documents these arguments through a sustained empirical investigation of the Vancouver 2010 Winter Olympics.

Plotting points on the Vancouver 2010 mega-event security-development-nexus

Between February 12th to 28th and March 12th to 21st, the Vancouver 2010 Olympics and Paralympics spanned more than 100 venues across the lower mainland of British Columbia. The cities of Vancouver, Richmond and the North-Shore Mountains (Whistler, Callaghan) hosted the largest peacetime security operation in Canadian history, with security expenditures reaching around $1 billion (all sums in Canadian dollars) (Lee 2009b). Aside from the enormous capital expenditures for planning and infrastructure that covered three security domains—the "theatre of operation" (Olympic venues), the "urban domain" in Vancouver as well as the mountainous Whistler area—the Vancouver 2010 Games triggered unprecedented security and intelligence alliances, foremost among them the Vancouver 2010-Integrated Security Unit (VISU), a federal initiative involving the RCMP, Canadian Security Intelligence Service (CSIS), the Canadian Forces, over 100 municipal police forces, 10 or more federal ministries, the Vancouver Olympic Organizing Committee (VANOC) and public transit agencies such as Translink Vancouver. New public–private networks between governments and the security industry, were also forged, by co-ordinating 7000 police from various forces across Canada, 4000 from the Canadian Forces (CF), and 5000 private security guards (Mercer 2009). Intelligence threat assessments were handled by a specialized agency, the Joint Information Group (JIG). The JIG, under the auspices of the RCMP, liaised with Canada's Integrated Threat Assessment Centre (ITAC), a national intelligence body bringing CSIS, local police forces and certain public agencies together to pool information and shape "actionable intelligence." Regulating all of these venues, agencies and organizations produced the networked intersections of tangled hierarchies and inter-jurisdictional modes of security governance set out in the next section.

The Vancouver 2010 MESDN: socio-economic restructuring and security governance

While the Olympics are hardly the tipping point in the transformation of urban spaces, the intensified expansion of capital that inevitably accompanies them leads to the displacement of certain groups and the gentrification of urban centres—which in turn generates contradictions, crisis tendencies and resistance

(Hiller 2006). Displacement has been a central feature of mega-events around the world. COHRE's investigation into gentrification and displacements unravels a string of human rights violations—72,000 people evicted from their homes in Seoul; 9,000 arrest citations issued to the homeless in Barcelona; temporary detention of homeless populations in Atlanta; and an incredible 1.5 million people displaced in the lead-up to the Beijing 2008 Games (COHRE 2007; Track 2009). For those who escape physical displacement, market price mechanisms in real estate, such as rent inflation and shrinking vacancy rates, serve the same function. As the Olympic economy stimulates infrastructure development in real estate and tourism, the surveillance measures now synonymous with property protection and public safety become even more prevalent, facilitating "an increasingly robust Security Protection (SP) market" (SIA 2007). The Polanyian "first movement" changes intensifying regimes of capital accumulation through regenerating urban spaces for infrastructural projects, then, have different effects on different populations. The "second movement" intensifies security and surveillance strategies to manage the contradictions, social ruptures and inequalities produced by the displacements.

In Vancouver, urban regeneration projects associated with the Olympics have been controversial since the earliest rumours of the city's bid (Shaw 2008). The most notable struggles revolved around the homeless population on the city's Downtown East Side (DTES). Citizens of the DTES live in the poorest postal area code in Canada—notably involving an open-air drug market—which is home to the largest population of people with HIV in North America. It has been reported that the DTES lost between 1085 and 1585 units of affordable housing since Vancouver was designated as Olympic hosts (COHRE 2007; IOCC 2010). VANOC's later promise to turn the athletes' village into affordable housing was consistently scaled back from thousands of units to 125 units (CBC 2010). Public unrest peaked in late 2008 when the global economy sank into recession. As an estimated 6 billion dollars was poured into Olympics venues, the provincial government announced drastic austerity measures, plunging many households into fiscal crisis (Mickleburgh 2010).

Social unrest over mega-events has typically motivated institutions of security governance to find ways to counter resistance and regulate, contain and/or incapacitate protests (Gillham and Noakes 2007). Today such expansion focuses on finding technical means to regulate "non-compliance" (Gill 1995). As mega-event hosts showcase their city to a global audience with the explicit aim of attracting tourists and capital investment, groups that threaten this image become impediments to commerce and therefore, in dominant discourses, to the prosperity and global reputation of the host city. Protecting the exclusive marketing deals already negotiated with particular corporations also requires greater surveillance. Thus, as Boyle and Haggerty (2009a) point out, the Olympics typically produce intensified efforts to "police the poor," "police commercial rights," and "police dissent" (58–72). The intersections between legal frameworks and security governance create the regulatory infrastructure deemed necessary to transform urban spaces into the "entrepreneurial city." These juridified relations, which legitimize ever more security and surveillance, reflect the articulation between legal regulation, security governance, and regimes of accumulation.

Policing the poor

Under the auspices of the Province of British Columbia's *Safe Streets Act* and *Trespass Act*, the Vancouver Police Department (VPD) adopted a strategy to "improve liveability by reducing disorder." Much like plans that emerged in Sydney and Atlanta, the policy states that "members will . . . . use existing legislation to specifically combat behaviour and activities that contribute to urban decay, including aggressive panhandling, squeegeeing, graffiti, public fighting, open-air drug markets, unlicensed street vending, the scavenger economy, and sleeping/ camping in city parks and other public spaces" (VPD Business Plan; Boyle and Haggerty 2009a: 61). The admitted purpose of the legislation was to expand the bureaucratic powers of the security regime to regulate "low-level misconduct." In the run-up to the Games, VPD's Beat Enforcement Teams (BETs) increased 200%. Another strategy designed to regulate homelessness, panhandling, and public disorder was *Project Civil City*, a Vancouver municipal programme that was first proposed in 2006. Supported by Vancouver's Business Improvement Associations (BIAs), and its Boards of Trade and Tourism, the programme's stated intention was to use the Olympics as a "catalyst" for reducing public disorder by granting extended discretionary powers to law enforcement. However, following resistance from Vancouver's robust activist movements, this proposal was abruptly scrapped in the lead-up to the Games. Reports of mass displacements and arrests of the poor during the Games were also constrained by political pressure (Vonn 2010).

This does not mean, however, that unprecedented security governance initiatives were derailed. VPD officers patrolled the Games with the Canada Border Services Agency's (CBSA) "immigration inland enforcement" agents (BCCLA 2010a) and Canadian Corrections Officers, for reasons that were never publicized. CBSA officers generally have the same powers as police officers to make arrests for immigration violations and deportations with their powers falling squarely under the Canadian Immigration and Refugee Protection Act. The CBSA now operates under the newly reconfigured Ministry of Public Safety and Emergency Preparedness, with orders to "ensure the twin goals of economic security and public safety," aptly exemplifying the nexus between security and development (Citizenship and Immigration Canada, 2004: 1; Gordon 2006: 126). Questions about the jurisdictional terms CBSA agents and immigration officers were operating under are still being debated, but one fact is incontrovertible: intensified low-level contact between these patrols allowed their respective agencies to collect and store personal information for future use. The potential for such information to be converted into electronic records known as "lookouts" (O'Connor 2006: 157) is worrying. Such "pop up" security regimes illustrate the Polanyian second movement in the context of one-off mega-events, and raise concerns about public accountability. To this day no justification for these arrangements has been presented (Vonn 2010).

Policing commercial rights

The sight of public and private security forces suppressing free speech to protect the exclusive marketing rights of Olympic sponsors has become commonplace. Although these developments legitimate public censorship to protect private interests (BCCLA 2010), the Olympic Charter authorizes granting exclusive marketing and distribution rights to particular sponsors in selected areas of the host city. Because governments are legally bound to protect these commercial interests, they can justify a range of intensified surveillance mechanisms and sanctions. Indeed, as Boyle and Haggerty (2009a: 63) point out, identifying violations of these commercial rights *presupposes* some sort of monitoring regime (emphasis ours). Private security officials acting on behalf of private, unelected corporations are thus empowered to conduct random searches, police citizens' consumption habits, record their personal information and remove citizens from Olympics venues for using or showing logos of "unapproved" brands (Klauser 2008). The 2000 Sydney Games, sponsored by Coca-Cola, saw private security guards searching event ticket-holders for "knives, weapons, or cans of Pepsi," and there were allegations that such laws were also used to crack down on buskers, political and religious demonstrations, individuals wearing anti-corporate logos, and panhandlers (Lenskyj 2008).

Although the legal structure was firmly in place (and remains so), no public controversies over enforcing commercial rights were reported at Vancouver 2010. Legislation that expanded the discretionary powers of the police to enter private homes to remove unapproved signage, following heavy public criticism, went largely unenforced—except for contestation over displays of public art (BCCLA 2009). Although authorities in Vancouver were savvy enough to avoid public embarrassment, this legally justified intensification of security has been successfully woven into the accumulation strategies pursued by the security industry, since host cities must prove they can provide a secure space before their application will be considered (Shaw 2008).

Policing dissent

Political resistance at mega-events is another global phenomenon, also one with distinctly local characteristics (Klauser 2008: 72–74; Shaw 2008; Lenskyj 2000, 2006). "Strategic incapacitation" has become a common policing strategy at mega-events. However, seeking to avoid criticism for the kind of overzealous policing displayed at the Copenhagen Climate Summit protests and the RNC convention in St. Paul, Minnesota, Vancouver authorities adopted a more tolerant, "negotiated management" style characterized by less visible forms of "social steering" (Gillham and Noakes 2007; della Porta and Reiter 1998).

Security officials originally planned to erect "optional" "free speech areas" near every competition venue, purportedly to provide a space for protests in busy areas around venues (IOCC 2010). After the BC Civil Liberties Association pointed out that these zones were in fact intended to accommodate the IOC and protect the Olympic brand and its sponsors (Vonn 2010), and under intensified media scrutiny, the plan for designated free speech areas was dropped.

However, the VISU policy of "infiltration" techniques (Graham 2009) that included plain-clothes police visits and phone calls to the homes and workplaces of activists (and their neighbours and co-workers) was not. The visits took place over a large area of BC, including Vancouver, Victoria, Kelowna and certain First Nations reserves. Reports also emerged of "unidentified police observing the public at City Hall, community meetings and educational events" (IOCC 2010). A letter from the Olympics Resistance Network (ORN) threatening legal action against the VISU if the visits did not cease was ineffective (IOCC 2010). In one instance, Victoria Police Chief Jamie Graham revealed that "a cop" was driving a coach full of activists to protest the Victoria Olympic Torch Relay, from Vancouver (Graham 2009). Such visits induced a chilling effect within activist circles. However, they also promoted solidarity and mobilization of dissent, by prompting activists to draw clear lines between "friend and enemy" (Anonymous local activist 2010). Such levels of resistance, plus the scrutiny of the world's media, prompted police to display considerable restraint at public events (such as protests at the Games and Olympic Torch Relay kick-off)—although it is alleged that this symbolic restraint was actually a strategy of "infiltration" where uniformed officers purposefully positioned themselves beside activists with backpacks during public protest (Graham 2009).

Strategies that "police the poor," "police commercial rights" and "police dissent" in the context of the mega-event security–development nexus have two related implications. First, they obscure the neoliberal foundations of the Olympic mega-event while simultaneously creating the necessary conditions for launching "pre-emptive" security of activists and dissidents. Mega-events, then, are less "zones of exception" than strategies that normalize accumulation, displace insecurity onto marginalized populations and the environment, and reproduce the social conditions necessary for capital accumulation. Democratic discussion on whether mega-events should continue socializing costs and privatizing profits are obscured by the "necessity" to provide intensified security governance at mega-events. Framing resistance to the Olympics in terms of security contributes to an architecture of power that avoids public debate over the direction of associated capital investments, and legitimizes a mass security apparatus that engages not just Olympic protesters but all (including environmental activists) who resist the accelerated development strategies associated with the Games.

The final section of the paper expands on a third element in the MESDN, security-specific accumulation regimes and the large-scale transformation of the public sector. The need for security at mega-events, we argue, is also expressed on the "supply side" through transnational relations of capital accumulation in the private security industry which use the MESDN to showcase its risk management strategies and products. Security governance is necessarily (though contingently) conditioned through informal industry norms and practices, coordination services firms, production alliances, subcontractor relationships, business associations, and private regimes (Cutler et al., 1999). Part of the pressure for "political-re-regulation"—the second movement in Polanyi's double movement—entails an expansion of security governance into regional and municipal/urban markets—where the identification of national security and infrastructure protection with commercial prosperity and accelerated development positions them to take advantage of the burgeoning urbanized "homeland-security-industrial complex" (Barry 2009).

Mega-event security accumulation regimes and tangled hierarchies of security governance

The global "recalibration of security" (Boyle and Haggerty 2009a) associated with the mega-event has made security strategies increasingly sub-national, regional, and urban in scale (Coaffee and Wood 2006). Security concerns are couched within an urban frame of reference, such that every security apprehension appears to be somehow urban and every urban issue is infused with security concerns. This would not be possible without the socio-economic and juridical restructuring of urban spaces of host cities outlined in section two, in concert with the direct power of transnational capital through network building and elite participation in the private security industry.

In purely economic terms, mega-events offer a bonanza of growth opportunities. The figures are staggering: in the 2000 Sydney Summer Olympics, approximately $180 million USD was spent on security operations. In the 2002 Salt Lake Winter Games, the first since 9/11, this jumped to almost $310 million USD, ballooning to an estimated $1.5 billion USD for the 2004 Athens Summer Olympics. Figures for the 2006 Torino Winter Games were $400 million USD. Varying estimates are provided for the 2008 Beijing Summer Olympics. The Chinese government puts official spending on security operations at $300 million USD, but more comprehensive estimates put this at $6.5 billion USD (SIA 2007; Boyle and Haggerty 2009a). The security budget for the 2010 Vancouver Winter Olympic Games, originally estimated to be $175 million CAD, ballooned to $900 million CAD by February 2009. The security cost of future Olympic Games, such as London 2012, is presently set at a staggering $1.5 billion GBP (Boyle and Haggerty 2009a).

Although mega-events serve as key sites of demand in the global security industrial complex, these security projects are in part an extension of the ballooning high-tech defence, military, intelligence and U.S. Department of Homeland Security (DHS) contracting industry. Out of the top-ten DHS contractors in 2008—Lockheed Martin, Northrup Grumman, IBM, L-3 Communications, Unisys, SAIC, Boeing, Booz Allen Hamilton, General Electric (GE), and Accenture—all have been involved in mega-event security projects.

The profits generated are enormous. As Boyle and Haggerty (2009a) point out, SAIC picked up $322 million USD profit from the Athens 2004 Games, although this is a mere fraction of their average annual revenues of $7–8 billion USD in recent years. The Beijing 2008 Olympics allowed various American firms (IBM, General Electric, Honeywell, and United Technologies) to expand into the Chinese domestic market (Bradsher 2007a; Haggerty and Boyle 2009a). The "Grand Beijing Safeguard Sphere" for example, launched high-tech surveillance systems in 600 Chinese cities (Bradsher 2007b). The Chinese association for surveillance studies predicts this market will top 43 billion USD by 2010 (Bradsher 2007c).

These massive infusions of economic capital offer rich opportunities for the accumulation of symbolic capital. As Bourdieu (1993) points out, control over material resources is closely linked to social and cultural legitimacy and relations of dominance. The overwhelming presence of the security apparatus at major public viewing events legitimates the enormous private sector role the security industry now plays. Opportunities to rethink and expand capital accumulation strategies abound—mega-events showcase "place-based" security technologies that can be marketed as "state of the art" systems to be used for homeland security. SAIC, the developers of security systems in Sydney 2000 and Salt Lake 2002, have successfully leveraged each mega-event to this end (Samatas 2007). Although corporations insist such contracts are more important "symbolically" rather than financially—Acklands-Grainger (AG), makers of systems that sniff out chemical, biological, nuclear and radiological threats, insists their biggest payoff will be increased visibility and influence in the emergency preparedness industry (Lee 2009a)—such claims should be viewed with scepticism.

Finally, the pivotal role played by security industry associations and their involvement in the revolving door that has developed between government, industry, and business must be highlighted. The British Security Industry Association (BSIA) has appointed its own Project Director for the 2012 Olympic Games, whose job is to provide "significant opportunities for UK business, particularly the security industry" (BSIA 2006). But the influence of the industry association is truly global. One network of international security trade associations, the Global Security Industry Alliance (GSIA), has member organizations in the U.S.A., China, Russia and Brazil, and over the last eight years each of these countries has competed for or is hosting an Olympics or World Cup mega-event. Security industry association involvement in mega-event planning desperately needs further empirical investigation.

Vancouver 2010 mega-event as key market for global security industry

The Vancouver 2010 Olympics also provided a key niche for expanding the global security industry market. With over $900 million CAD at stake, a number of government agencies negotiated contracts with private security firms.

One of the largest was Honeywell, given a contract worth $35.5 million to provide a Perimeter Intrusion Detection System (PIDS). The PIDS integrated a range of technologies (surveillance cameras, motion sensors, ground- and water-based radar) into a single system and promised "intelligent video" capabilities that could distinguish "normal activity" from "that which poses a security risk" (Honeywell Critical Infrastructure 2009). SAIC was a silent partner on the Honeywell contract. However, one VANOC source has suggested that the PIDS system did not function properly and experienced significant cost overruns (Mackin 2010), reminiscent of Athens 2004 where the state-of-the-art SAIC C4I system failed (Samatas 2007). Other key contracts included one to Garret Metal Detectors for 550 walk-through metal detectors and 1,100 hand-held detector devices—issued despite reports from a previous Olympic organizing executive that airport style checks provide "illusory security" because they cannot detect non-metallic materials in explosives (Boyle and Haggerty 2009a: 50).

Vancouver 2010 also purchased a public order technology, the Long Range Acoustic Device (LRAD), developed by the American Technology Corporation for civilian crowd control (Canadian Press 2009). LRADs use a high-pitched frequency sound to break up crowds; a piercing high-pitched noise so powerful can potentially cause permanent eardrum and heart damage. Because it is classified as a "device," not a "weapon," its manufacturers can circumvent US trade embargoes banning weapon sales to particular countries—a loophole that allowed the company to close lucrative deals in Beijing in 2008 (Hambling 2008). Previous acquisition of an LRAD by the Pittsburgh and Toronto Police in advance of their G20 Summit responsibilities indicates this will be a central technological feature at future mega-events.

Contemporary Security Canada (CSC) received a $97.42 million dollar contract to secure 5,000 private security officers for Vancouver 2010, more than in the entire province of British Columbia at the time (CBC 2009). CSC performed background checks and fingerprinting of applicants but a shortened training programme led privacy critics and local competitors to cry foul—but no embarrassing difficulties emerged during the Games (CBC 2009).

Intensified integration of police-related and military-related activities characterize the MESDN, and Vancouver 2010 was no exception. Before the Olympics started, the Vancouver Police Department sought new recruits with "experience in the military or [those who have] spent time in war zones in non-military roles" (Hogben 2008). This raises questions about the impact such a diffusion across institutional boundaries allows, perhaps permitting a reshaping of "law and order" strategies in the absence of case law or explicit institutional coordination. But the military was directly involved as well: the Games were secured with a series of portable command posts and F-16 fighter jets. The Department of National Defence involved tendered contracts worth $1.5 million for partial renovation of an Air Canada hangar at the Vancouver International Airport (YVR) to bring it up to mega-event capabilities—another legacy of the Games.

Reconfiguring networks of security governance

Inter-agency coordination has been the hallmark of security governance at the Vancouver 2010 Olympics. Intensive capital investment in security has spurred significant transformations in security and public safety agencies at all levels of government. The capital-intensive security projects and political regulation associated with mega-events today has led to intensive capital investment in the public sector, expressed by integrating security and surveillance organizations and technological systems (Boyle and Haggerty 2009a).

Such systems rely on multiple forms of coordination across different organizational configurations, policy regimes and technological systems. As policy learning from previous mega-events takes place, public security governance is restructured to facilitate integration. Providing a security blanket for hundreds of thousands of citizens spanning across dozens of venues and several kilometres of disparate geographical spaces requires the public sector to establish a massive, integrated organizational structure. Associated with this large-scale transformation of the public sector is the urgent "need" to project the right image: venues must be seen as capable of delivering a "safe" mega-event. Thus, managing the reputation of the host city or country is a further cultural factor spurring growth of the MESDN (Coaffee and Rogers 2008).

Mega-event security and "reputational risk"

Managing reputations through successful events is now seen as evidence of a city's stability, financial success and global identity (Hall 2006; Xu 2006). Demonstrating security competence in spaces of commercialism and leisure elevates a city's "brand," its competitive edge in the global marketplace of urban centres, and it is "a vital selling point in [a] global city 'offer'" (Coaffee and Murakami Wood 2006: 508). Senior Vancouver 2010 officials and police repeatedly mentioned that one of their most important objectives was to avoid disruption and thereby maintain/burnish Vancouver's reputation. But to be successful, the security apparatus had to maintain low levels of visibility. Highly visible security in consumptive spaces creates a culture of fear which threatens one of the core commercial purposes of the Olympics, the promotion of consumption and tourism. As Bud Mercer, Commander of VISU, consistently framed it, the Olympics must "remain about sports, culture, and sustainability, and not about security" (Mercer 2009). Chief Jamie Graham, former head of Olympic planning for the Vancouver Police Department, delivered a similar message: "you never want your community remembered as a community that couldn't protect something," adding, "I don't care what it takes, I don't care what it costs, it [a disruption] is not going to happen in Vancouver" (Graham 2009). Efficient inter-agency coordination is key to both invisibility and effectiveness in security management.

Security governance is also crucial in burnishing the reputations of security professionals and firms on a world stage, and it functions as a strategic asset in strategies of accumulation. Coaffee and Rogers sum it up nicely: "Resilience, safety, and security have . . . become an . . . important tool in the armoury of reputation managers and place promoters as security, marketing, economic development and regeneration have become necessarily intertwined" (2008: 215).

However, it is not just security agencies and elite coalitions who valued the chance to promote Vancouver as a secure place for investment (Surborg et al., 2008). Global reputation and image making are but one component of the large-scale transformations that take place in the public sector in the context of mega-events—the strategic meso-level transformations of security governance forged through and necessitated by organizational and technological integration are more significant still.

Security governance in the context of Vancouver 2010

Inter-agency coordination in Vancouver was structured through the Major Events Template (MET). Fashioned from the Incident Command System, the MET is a crisis management tool designed to establish a series of positions, roles and responsibilities for public and private sector agencies (Boyle and Haggerty 2009b). Intended as a policy template applicable to all large-scale security projects, the MET has already been deployed in subsequent mega-events in Canada, notably the G8/20 Summit in Toronto. The inter-agency relationships (and contacts and technologies) forged in Vancouver are already bearing fruit. An important empirical question, then, is whether, and to what extent public participation is factored into the MET template. At this point it appears to be limited or absent: security professionals worldwide have formed an epistemic community, a closed circle where specialized knowledges spanning multiple individuals, groups, stakeholders and organizations circulate (Haas 1992; Klauser 2008 69–70; Boyle this volume). This closed circle has been reinforced in the IOC itself—the IOC security consultant and former head of Sydney 2000 security, for example, did consulting jobs for VANOC. Much like the revolving door between government and industry in the broader homeland-security-industrial complex, mega-event Olympic security blurs the boundary between the public–private and the civilian–military divide. "Proto-institutional" relationships (Lawrence et al. 2002) that expand the reach and capital of the security industry are developed, penetrating into national, provincial/state-level, and municipal government networks, networks that persist and reproduce long after the mega-event has ended. Through these path-dependent, path-shaping neoliberal disciplinary mechanisms (Gill 2003), "mega-event security epistemic communities," interdependent networks of a global–commercialized and urban–militarized nature, are reproduced throughout the public sector.

Mega-events provide both the justification and the capital to integrate relatively disparate high-tech surveillance appliances into comprehensive surveillance networks such as the MET. Technological integration builds social and jurisdictional inter-organizational relationships, as security firms interact to integrate disparate high-tech appliances into centrally managed technological systems. For instance, the high-tech security apparatus provided by SAIC–Siemens at the Athens 2004 Games, the C4I system providing command, control, communications and integration, was perhaps the most extensive in mega-event history. C4I integrated a network of 29 subsystems into one single command system that linked Greek police, firefighters, coastguards, and armed forces (Samatas 2007: 221). The system also boasted 130 fixed and 5 mobile command centres, vehicle tracking devices, underwater motion sensors, 1250 to 1600 surveillance cameras, and a surveillance blimp—all underpinned by a 7,000-strong Greek security force (Samatas 2007: 221). Integration spanned across 35 Olympic venues in Athens as well as four other cities, linking critical infrastructure facilities, including power stations, water utilities and oil and gas reservoirs (Samatas 2007: 221). This scale of integration provided a publicly funded opportunity for firms to assess their systems and for purchasers to reallocate uses of surveillance technologies perimeter detection systems, surveillance cameras and informational databases, as "durable assets" for redeployment well beyond their original application (Chase 2010). And the fact that the C4I system failed (!) is apparently insignificant, a multi-billion dollar blip, a mere "learning opportunity" for the security industry. Public participation and debate over the design and wisdom of adopting such far-reaching—and unreliable—systems is virtually non-existent, only dramatically inflating costs have attracted sustained critical comment.

Several private firms position integration practices at the core of their corporate security strategies (e.g., Honeywell Security 2009). As one security analyst states, "unified security platforms that seamlessly blend all security systems into one are the holy grail of the industry" (Himmelsbach 2009). A director of AFI International, a Canadian security firm that specializes in securing corporate events and labour disputes (both smaller-scale manifestations of event-led security processes), thinks budget issues alone will force integration (Cummer 2009). Global economic pressures are changing the strategic calculations of the security industry—"[g]one are the days when a company wants a security expert . . . they [now] want a business expert with a strong focus on security" (Cummer 2009). The budget constraints compelled by the most recent contraction of the global economy have reinforced the imperative to integrate all organizational and technological infrastructures into multi-tiered and multi-scalar arrangements.

Conclusion

The strategic and structural coupling between regimes of accumulation (Aglietta 1979) and modes of governance (cf. Jessop 1995) hinges on a mega-event specific security–development nexus that we suggest occurs in three main ways. First, security governance at mega-events is the liberal fulcrum which trumps dissent whenever disparities in public spending and the displacement of externalities on populations and the environment are contested. The emergence of security governance regimes at mega-events is expressed in a Polanyian (second) movement as a contingent, though necessary, arrangement to secure accelerated development through strategies of political re-regulation such as "policing the poor," "policing dissent," and "policing commercial rights" (Boyle and Haggerty 2009a). Security is not supplemental, it is constituted through capital imperatives associated with hosting a successful, incident-free event. Intersections between legal frameworks and security governance are the regulatory means for transforming urban spaces into "entrepreneurial cities," forming a set of social relations which the mega-event inevitably grafts onto and deepens. These juridified relations provide the basis for legitimating ever more claims to enlarge security and surveillance.

Shifting from urban and local scales, we articulated a second major dimension to the MESDN. Here, the real and perceived necessity of security governance at mega events is expressed on the "supply side" through transnational relations of capital accumulation in the private security industry. Security governance is necessarily (though contingently) conditioned through a myriad of industry norms and practices, firms offering coordination services, production alliances, subcontractor relationships, and business associations. Heavy state-subsidized capital investment triggers key meso-level transformations in governance regimes specializing in social ordering practices, offers huge political and economic opportunities for actors selling security services and technology, and gives public sector agencies the opportunity (often sold as the obligation) to "grow" their emergency preparedness, public safety and security infrastructure.

This institutional restructuring of key areas of the public sector should lead researchers to consider the role of economics and business models of efficiency as factors shaping the development and application of surveillance practices. It should also prompt researchers to privilege the importance of meso-level explanations of institutional integration at mega-events. With more and more mega-events located in "the global south"—countries such as Brazil, India, and South Africa—unpacking the MESDN must become more than ever a theoretical, methodological, and empirical priority.

References

Aglietta, M. 1979. *A Theory of Capitalist Regulation: The U.S. Experience*. London: NLB.
Anonymous local activist. 2010. Personal Communication. May 12.
Bajc, V. 2007. Surveillance in public rituals: Security meta-ritual and the 2005 U.S. presidential inauguration. *American Behavioral Scientist* 50 (12): 1648–1673.
Barry, T. 2009. *The national security complex—integrating military, intelligence, and homeland security*. URL: http://borderlinesblog.blogspot.com/2009/09/national-security-complex-integrating.html (web blog accessed May 19, 2010).
BCCLA (British Columbia Civil Liberties Association). 2009. City orders anti-Olympic mural removed from Gallery. URL: http://www.bccla.org/pressreleases/09mural.html (accessed May 12, 2010).
—— 2010. "Free speech zone" crowded with pro-Olympic displays. February 9. URL: http://www.bccla.org/pressreleases/10Free_speech_zone.html (accessed May 12, 2010).
—— 2010a. "2010–02–25- CBSA 'Inland Enforcement.'" Video URL: http://vimeo.com/9751444 (accessed May 19, 2010).

Bigo, D. 2005. Globalized (in)security: the Field and the Ban-opticon. In *Terror, Insecurity and Liberty: Illiberal Practices of Liberal Regimes after 9/11*, Didier Bigo and Anastassia Tsoukala (eds), pp. 10–48. New York: Routledge.

Bourdieu, Pierre. 1993. *Sociology in Question*. London: Sage.

Boyle, P. 2011. Knowledge networks: Mega-events and security expertise. In *Security Games*, Colin Bennett and Kevin D. Haggerty (eds). London: Routledge.

Boyle, P., and K. Haggerty. 2009a. Spectacular security: Mega-events and the security complex. *International Political Sociology* 3 (3): 257–274.

—— 2009b. Privacy Games: The Vancouver Olympics, Privacy and Surveillance. Ottawa: A Report to the Office of the Privacy Commissioner of Canada.

Bradsher, K. 2007a. China finds American allies for security. *New York Times*. December 28. URL: http://www.nytimes.com/2007/12/28/business/worldbusiness/28security.html (accessed December 12, 2009).

—— 2007b. China enacting high-tech plan to track people. *New York Times*, August 12, A1. *New York Times*, A1. URL: http://www.nytimes.com/2007/08/12/business/world business/12security.html (accessed December 12, 2009).

—— 2007c. An opportunity for Wall Street in China's surveillance boom. *New York Times*, September 10, A1. URL: http://www.nytimes.com/2007/09/11/business/world business/11iht-11security.7457945.html (accessed December 12, 2009).

Brighenti, M. 2008. Democracy and its visibilities. In *Surveillance and Democracy*, Kevin D. Haggerty and Minas Samatas (eds), pp. 51–68. New York: Routledge.

BSIA (British Security Industry Association) Report 2006. Olympic Games 2012 Project Director appointed. December 7. URL: http://www.bsia.co.uk/aboutbsia/news/M954SD10180 (accessed May 20, 2010).

Canadian Press. 2009. Vancouver police's new weapon could pose risk. November 11. URL: http://www.ctvbc.ctv.ca/servlet/an/local/CTVNews/20091111/sonic_weapon_09 1109/20091111/?hub=BritishColumbiaHome (accessed May 20, 2010).

CBC. 2009. 2010 Olympics worth $100M to private security firms. April 17. URL: http://www.cbc.ca/sports/amateur/story/2009/04/17/sp-olympics-security-2010.html (accessed November 13, 2009).

CBC News. 2010. Vancouver cuts Olympic Village social housing. April 23. URL: http://www.cbc.ca/canada/british-columbia/story/2010/04/23/bc-olympic-village-social-housing.html (accessed May 1, 2010).

Chase, S. 2010. Summit security costs put federal priorities under microscope. *The Globe and Mail*. May 26. URL: http://www.theglobeandmail.com/news/world/g8-g20/summit-security-costs-put-federal-priorities-under-microscope/article1582266/ (accessed May 26, 2010).

Citizenship and Immigration Canada. 2004. Creation of the Canada Border Services Agency. Ottawa: Citizenship and Immigration Canada.

Coaffee, J., and P. Rogers. 2008. Reputational risk and resiliency: The branding of security in place-making. *Place Branding and Public Diplomacy* 4 (3): 205–217.

Coaffee, J., and D. M. Wood. 2006. Security is coming home: Rethinking scale and constructing resilience in the global urban response to terrorist risk. International Relations 20 (4): 503–517.

COHRE The Centre on Housing Rights and Evictions. 2007. Fair Play for Housing Rights: Mega-Events, Olympic Games and Housing Rights. Geneva, Switzerland. URL: http://www.cohre.org/sites/default/files/fair_play_for_housing_rights_2007_0.pdf (accessed November 30, 2010).

Cummer, L. 2009. Can integration solve all our security problems? *Canadian Security Magazine*. June. URL: http://www.canadiansecuritymag.com/Integration/News/Can-integration-solve-all-our-security-problems.html (accessed May 20, 2010).

Cutler, A. C., V. Haufler, and T. Porter (eds). 1999. *Private Authority and International Affairs*. New York: SUNY Press.

della Porta, D., and H. Reiter. 1998. Policing Protest: The Control of Mass Demonstrations in Western Democracies. Minneapolis: University of Minnesota Press.

Escobar, A. 1995. *Encountering Development: The Making and Unmaking of the Third World*. Princeton, NJ: Princeton University Press.

Gaffney, C. 2010. Mega-events and socio-spatial dynamics in Rio de Janeiro, 1919–2016. *Journal of Latin American Geography* 9 (1): 7–29.

Gill, S. 1995. Globalisation, market civilisation, and disciplinary neoliberalism. *Millennium: Journal of International Studies* 24 (3): 399–423.

—— 2003. *Power and Resistance in the New World Order*. New York: Palgrave Macmillan.

Gillham, P. F., and J. A. Noakes. 2007. "More than a march in a circle": Transgressive protests and the limits of negotiated management. *Mobilization: An International Quarterly* 12 (4): 341–357.

Gordon, T. 2006. *Cops, Crime and Capitalism: The Law-and-Order Agenda in Canada*. Halifax: Fernwood Books.

Graham, J. 2009. Best Practices in Shared Jurisdictional Environments Where Multiple Police, Fire, and Ambulance Services Work in Concert to Promote Public Safety. Talk Presented at the Reboot Vancouver International Security Conference, November 30, Vancouver.

GSIA (Global Security Industry Alliance). 2009. Website. URL: http://www.siaonline.org/gsia.aspx?id=1488&linkidentifier=id&itemid=1488 (accessed May 20, 2010).

Haas, P. M. 1992. Introduction: Epistemic communities and international policy coordination. *International Organization* 46 (1): 1–35.

Hall, C. M. 2006. Urban entrepreneurship, corporate interests and sports mega-events: The hard outcomes of neoliberalism. In *Sports Mega-Events: Social Scientific Analyses of a Global Phenomenon*, J. Horne and W. Manzenreiter (eds), pp 59–70. Oxford: Blackwell Publishing.

Hambling, D. 2008. US "Sonic Blasters" Sold To China. *Wired Magazine*, May 15. URL: http://www.wired.com/dangerroom/2008/05/us-sonic-blaste/ (accessed May 20, 2010).

Harvey, D. 1989. *The Condition of Postmodernity: An Enquiry into the Origins of Cultural Change*. London: Wiley-Blackwell.

Hettne, B. 2010. Development and security: Origins and future. *Security Dialogue* 41 (1): 31–52.

Hiller, H. H. 2006. Post-event outcomes and the post-modern turn: The Olympics and urban transformations. *European Sport Management Quarterly* 6 (4): 317–32.

Himmelsbach, V. 2009. Considering unified security vs. best-of breed. *Canadian Security Magazine*. September 11. URL: http://www.canadiansecuritymag.com/Integration/News/Considering-unified-security-vs.-best-of-breed.html (accessed May 20, 2010).

Hogben, D. 2008. Vancouver police hire former mercenaries for 2010. *Vancouver Sun*. September 12.

Honeywell Security. 2009. URL: https://www.honeywellintegrated.com/ss/index.html (accessed October 28, 2009).

IOCC (Impact On Communities Coalition). 2010. Olympic Oversight Interim Report Card 3. February 25. URL: http://iocc.ca/documents/2010-02-25_IOCC_3rdInterim Report Card.pdf (accessed May 20, 2010).
ITAC (Integrated Threat Assessment Centre). 2008. 2010 *Vancouver Winter Olympics: Terrorist Threat to Vancouver Area Facilities*. Ottawa: Integrated Threat Assessment Centre.
Jessop, B. 1995. The regulation approach, governance and post-Fordism: Alternative perspectives on economic and political change? *Economy and Society* 24 (3): 307–33.
—— 2002. Capitalism, the regulation approach, and critical realism. In *Critical Realism and Marxism*, Andrew Brown, Steve Fleetwood and John Michael Roberts (eds), pp. 88–115. New York: Routledge.
Klauser F. 2008. Spatial Articulations of Surveillance at the FIFA World Cup 2006™ in Germany. In *Technologies of Insecurity*, K. Franko Aas, Helene Oppen Gundhus, and Heidi Mork Lomell (eds), pp. 61–80. London: Routledge.
Krahmann, E. 2003. Conceptualizing security governance. *Cooperation and Conflict* 38 (1): 5–26.
Lawrence, T. B., C. Hardy, and N. Phillips. 2002. Institutional effects of interorganizational collaboration: The emergence of proto-institutions. *Academy of Management Journal* 45 (1) 281–290.
Lee, J. 2009a. Why the 2010 Olympics security budget is still secret. *Vancouver Sun* February 10. URL: http://communities.canada.com/vancouversun/blogs/insideolympics/archive/2009/02/10/why-the-2010-olympic-security-budget-is-still-secret.aspx. (accessed February 12, 2009).
Lee, J. 2009b. Heavyweight security eyed. *Vancouver Sun*. October 29.
Lenskyj, H. J. 2000. *Inside the Olympic Industry: Power, Politics, and Activism*. New York: State University of New York Press.
—— 2008. *Olympic Industry Resistance: Challenging Olympic Power and Propaganda*. New York: State University of New York Press.
Mackin, B. 2010. Bob Mackin, 24 News, Personal Communication. November 13, 2009.
Mercer, B. 2009. Presentation delivered at Reboot Vancouver International Security Conference, November 30, 2009. Vancouver.
Mickleburgh, R. 2010. Olympics fail to win over big chunk of B.C.: Poll B.C. Premier Gordon Campbell may be disappointed if he is hoping for a Games-based popularity bounce. *The Globe and Mail*. March 12. URL: http://m.theglobeandmail.com/news/national/british-columbia/olympics-fail-to-win-over-big-chunk-of-bc-poll/article1496992/?service=mobile (accessed May 13, 2010).
Newton, C. 2009. The reverse side of the medal: About the 2010 FIFA World Cup and the beautification of the N2 in Cape Town. *Urban Forum* 20 (1): 93–108.
O'Connor, D. 2006. A New Review Mechanism for the RCMP's National Security Activities. Commission of Inquiry into the Actions of Canadian Officials in Relation to Maher Arar. Ottawa.
Peck, J., and A. Tickell. 2002. Neoliberalizing space. *Antipode* 34 (3): 380–404.
Pieterse, J. N. 2000. After post-development. *Third World Quarterly* 21 (2): 175–191.
Polanyi, K. 2001. *The Great Transformation: The Political and Economic Origins of Our Time*. Boston: Beacon Press.
Roche, M. 2000. *Mega-events and Modernity*. London: Routledge.
Samatas, M. 2007. Security and surveillance in the Athens 2004 Olympics: Some lessons from a troubled story. *International Criminal Justice Review* 17 (3): 220–238.

Security Industry Association (SIA). 2007. China Security Market Report: Special Supplement, Olympic Update. Shanghai 2010 World Expo.

Shaw, C. A. 2008. *Five Ring Circus: Myths and Realities of the Olympic Games*. Gabriola Island, BC: New Society Publishers.

Stern, M., and J. Ojendal. 2010. Mapping the security–development nexus: Conflict, complexity, cacophony, convergence? *Security Dialogue 41* (1): 5–29.

Surborg, B., R. Van Wynsberghe, and E. Wyly. 2008. Mapping the Olympic growth machine. *City* 12 (3): 341–355.

Track, L. 2009. Downtown Eastside residents lose out in the 2010 Olympics. *Georgia Straight*. February 26. URL: http://www.straight.com/article-203867/laura-track-downtown-eastside-residents-lose-out-2010-olympics (accessed May 20, 2010).

Vonn, M. 2010. Personal communication. May 16.

VPD. 2009. *Vancouver Police Department 2009 Annual Business Plan*. Vancouver: Vancouver Police Department.

Xu, X. 2006. Modernizing China in the Olympic spotlight: China's national identity and the 2008 Beijing Olympiad. In *Sport Mega-events: Social Scientific Analyses of a Global Phenomenon*, J. Horne, and W. Manzenreiter (eds), pp. 90–107. Oxford: Blackwell Publishing.

Zedner, L. 2003. Too much security? *International Journal of the Sociology of Law* 31 (3): 155–184.

Chapter 10

Knowledge networks
Mega-events and security expertise

Philip Boyle

In responding to a question from the media about security preparations for the 2010 Vancouver Winter Olympics, a spokesperson for the RCMP, Canada's federal police force and lead security agency for the Games, made the following comment: "We look a lot to other countries as well, where events have happened. That teaches us a little bit about what security was put in place [and] what kind of problems they had" (Thompson 2009: 3). This seemingly innocuous postscript, offered almost as an afterthought, is a singular yet telling statement that touches on a broader set of activities wherein authorities involved in major event security learn—or publicly *profess* to learn—from the experience of those managing comparable events in other countries. These processes are in keeping with the efforts of event governing bodies to make the accumulated experience or "know-how" of past events available to future hosts through formal knowledge transfer programs, which in turn is in keeping with the wider embrace of knowledge management amongst global governing institutions (Ilcan and Phillips 2008).

These processes of teaching, learning, and emulation are not new to the field of major events or to policing in general. Adapting to the failures of the past, either in an organization's own history or that of others, is one way that policing organizations change over time, even if that change is slow, uneven, or externally driven (Waddington 2008). In the field of major events, informal fact-finding and observation programs arranged between police agencies involved in similar events is a "well-honored tradition in the event security community," according to one prominent US event security planner (Bellavita 2007: 20). This chapter argues, however, that events of the preceding decade have dramatically intensified and expanded these processes of inter-event scrutiny amongst public policing agencies while simultaneously drawing non-governmental actors into the fold. In Durkheimian terms, this "tradition" is rapidly becoming a formally differentiated and functionally specialized field of security expertise that now approximates what others might call transnational epistemic networks (Haas 1992). Emerging from the interrelated issues of the fragmentation of the state monopoly on security provision, the increasing prominence of non-state governance organizations and private sector actors in security governance, and the recurring problem of how to secure large, complex, and temporary public events from assorted risks, this field includes state and local law enforcement bodies, public safety and intelligence agencies, international sporting federations, international governance organizations, and security consultancy and technology firms along with a host of mediating actors including event management and logistics firms, industry associations, and public policy think tanks. The dense but shifting linkages between these various actors collectively constitute what I refer to in this chapter as *security knowledge networks*, which not only facilitate the movement of event-specific security expertise between geographically and temporally distant locales, but also serve as key channels through which contemporary security rationalities and technologies are consolidated and disseminated globally.

This chapter maps out these security networks and assesses their impact. Conceptual inspiration for this analysis derives in part from the criminological literature on security networks (Johnston and Shearing 2003; Dupont 2004), which has yet to empirically examine transnational security networks. I also draw upon the comparative public policy studies concerned with policy transfer and convergence (Haas 1992; Dolowitz and Marsh 1996). This literature is useful for concepts such as lesson drawing, emulation, elite networking, and epistemic communities, yet implicit in this perspective is the tendency to connote that policy makers engage in rationalistic and voluntaristic processes of evaluation and adoption (Stone 2004). For example, in his early and influential discussion of lesson drawing Rose (1991: 4) states that "if the lesson is positive, a policy that works is transferred, with suitable adaptations. If it is negative, observers learn what not to do from watching the mistakes of others." Rose concedes that "lesson drawing is part of a contested political process" (1991: 6), yet politics remains secondary to the rationalistic actions of policy makers, which "tends to be voluntaristic" (1991: 9). In contrast, more contemporary authors have sought to reconceptualize this process as being inherently socially constructed and situated within fields of power and politics (Stone 2004; Ilcan and Phillips 2008; McCann 2008; Peck and Theodore 2010). Rather than starting with the view that policy knowledge circulates in flat, apolitical space in accordance with the rational intentions of policy makers, this literature uses concepts such as mobility, mutation, and nonlinearity to accentuate that power and politics structure the market for ideas and best practices, "which is rarely, if ever, just about transferring policy knowledge and technology from A to B" (Peck and Theodore 2010: 169–170).

This chapter is therefore concerned less with the "success" and "failure" of these networks in transferring knowledge as it is with two main issues pertaining to contemporary security governance. The first is how different forms of expertise are mobilized around specific security issues (Klauser 2009); in this case, the "moving target" of the Olympic Games. In doing so, this chapter avoids a technologically deterministic account of major event security by examining the knowledge infrastructures that connect these governmental domains across time and space. The second aim of this chapter is to consider the broader significance of these knowledge networks for individuals and societies. Of particular interest here is how these networks are instrumental in consolidating and promoting the wider objectives of major event security, objectives that involve positioning the Games as accelerants for the production of long-term security governance legacies for the host city and country. This extends the significance of these networks beyond the issue-specific domain of major events to include their role as important pedagogical vehicles for the "making up" and globalization of security expertise in the post-9/11 era. This pedagogical function is directed particularly towards cities of the global south where major events are being held with increasingly regularity and where the legacy building potential is much greater. As such, these knowledge networks provide not only an opportunity to examine how different forms of expertise are assembled around particular security issues, but also to analyze one of the key channels through which contemporary security expertise itself is constituted and globalized.

This chapter draws on data collected as part of a broader project initiated in 2006 researching the security, surveillance, and policing aspects of the Olympic Games in the post-9/11 period. To date this research has entailed four research trips to Vancouver, British Columbia, one of which occurred during the 2010 Winter Olympic Games, one to Ottawa, Ontario, and one to London (UK) to conduct interviews and attend a security industry conference on major events. Twenty-eight semi-structured interviews have been conducted in total for this research, thirteen of which are drawn upon for this chapter. These interviews include representatives from the RCMP, local and provincial public safety officials in British Columbia, current and former United Nations officials, and a variety of individuals from private industry. Where possible these interviews were conducted in person but six of thirteen were conducted by phone. This research has also amassed a small archive of wide-ranging government and non-governmental documents dealing with mega-event security, including post-event analysis and recommendation reports from government and non-governmental observers, testimonials, speeches, and published interviews with individuals involved with major events, law enforcement manuals and trade journals, and security industry publications and association reports. It also uses original documents regarding Canada's preparations for the 2010 Games acquired through Freedom of Information requests to the RCMP, Public Safety, and other relevant government agencies. The project also draws on publicly available news sources and on the existing academic literature on the Olympics and other major events.

Security knowledge assemblages

Dupont (2004: 79) notes that "security networks are porous and tracing boundaries can be a perilous exercise." This is particularly true in the context of the Olympics where multiple knowledge networks intersect, overlap, and are continually shifting in step with the march of the Olympic industry. This plurality and fluidity renders any attempt to map these networks unavoidably partial. Nonetheless, three key institutional networks can be identified: state institutional networks, transnational networks and non-state institutional networks. Each has the explicit purpose of pooling resources in order to facilitate the transfer and diffusion of practices across an institution (Dupont 2004: 80). Private security firms, while not constituting networks per se, are also significant transfer agents (Stone 2004).

State institutional networks

The approach taken by the USA and Canada towards major events is illustrative of state institutional networks. Both countries have major event designations that, when invoked by the relevant authorities, signal that the event has been deemed to require exceptional security measures and triggers the involvement of a predefined network of state agencies. In the USA this designation is called a National Special Security Event (NSSE). Once this designation is invoked by the President or the Director of Homeland Security, the US Secret Service (USSS) is appointed as the lead federal agency for developing and implementing security plans in conjunction with state and local authorities, the Federal Bureau of Investigation (FBI) in charge of intelligence and threat assessments, and the Federal Emergency Management Agency (FEMA) in charge of consequence management (Reese 2008). Canada takes a comparable approach. Canada's Minister of Public Safety has authority to declare any event a "major event" if it is deemed to be of national or international significance and where the overall responsibility for security rests with the federal government of Canada. This designation appoints the RCMP as the lead security planning and coordination agency in conjunction with other federal agencies, provincial public safety and emergency management authorities, and local law enforcement bodies.

The USSS and RCMP each have internal units responsible for major event security policy within their respective countries; in the USA the Major Events Division and in Canada the Major Events and Protective Services Division. These specialized institutional units fulfill a multitude of roles including serving as central liaison points for different policy divisions within the agency, coordinating planning efforts between headquarters and operational units, supporting operational planning, and assisting to develop unified command models for each specific event. Key to the latter function is the use of organizational templates that further specify how law enforcement, public safety, and intelligence agencies should (ideally) interact, define command authority and lines of communication, specify roles and responsibilities for middle- and upper-level planners, and set out timetables and milestones for operational planning. In both countries these templates are based on the Incident Command System (ICS), an emergency management organizational structure that developed out of wildfire management in California in the 1970s that is now used across North America.

These templates, which in Canada is little more than a large binder complete with organizational charts, checklists, timelines, job descriptions, examples of useful forms, and numerous technical appendices, provide standardized yet evolving starting points for major event security planning within each country.

As the Director of the RCMP Major Events and Protective Services explained in an interview, the template "standardizes our system so that someone who is just starting to plan can look at this and get a sense of where to start and who to talk to," yet it is "always evolving because with every event the lessons learned and the best practices are reintegrated so that it helps us evolve our practices and processes." It should be noted that these templates are primarily organizational tools pertaining to command structures, communications, and inter-agency working relationships, with operational planning falling to event-specific units. Nonetheless, these divisions play a crucial role in maintaining continuity between past and future events at the policy level by acting as institutional memory banks for best practices and lessons learned from past events which are codified in a formalized template that will shape how future events are managed in the country. As such, these units function as core nodes of state institutional networks insofar as they are "repositories for resources and information that each member of the network can access" (Dupont 2004: 80) that facilitate the mobility of knowledge within each respective agency. As the Director of the RCMP Major Events and Protective Services Division explained in an interview, "knowledge is reported through appropriate channels within the force and maintained, and that's what the Major Events Division is doing and has done in the past. We manage information on a needs basis."

Learning from others: transnational networking and non-state institutional networks

As the quote from the RCMP officer at the beginning of this chapter suggests, preparations for the 2010 Games included observing major events in other countries. As such, these preparations provide an entry point into transnational knowledge networks. Interactions with other countries is necessarily more complicated than facilitating knowledge mobility within a single institution; in Canada's case, these international activities have been coordinated in large part through the Office of the Coordinator for 2010 Olympics and G8 Security, which acts as a network "hub or switch that centralizes all outgoing flows from national sub-units and despatches [sic] back all incoming flows" (Dupont 2004: 81).

Canadian officials have been invited observers at all Olympics since the 2002 Games. Such activities are, as one security planner involved with the 2006 Turin Games put it in an interview, "a regular occurrence for large-scale events." Depending on the relationship between the countries involved, these observation programs can involve small groups of planners sitting in on a few meetings and debriefing sessions, or fully embedded planners shadowing counterparts in task-specific domains for weeks at a time. Longer, more in-depth observations allow the exchange of "plans and lessons learned and documents and things like that" but also in "building that informal relationship," according to another planner involved in the 2002 Games, that is important to effective learning in general. The densest connections, however, have been between Canada and the USA.

Representatives from the RCMP, armed forces, and public safety and emergency management agencies in Canada have attended and observed security planning and operations for the 2002 Winter Games, a number of Super Bowls, and a series of other NSSEs in the years preceding the 2010 Games. These activities are complemented by a range of professional conferences and working seminars hosted by US authorities which RCMP personnel have "attended, reviewed, and [also where they have] shared best practices" (RCMP 2005: 3), including a post-Games review of the USSS plans and after-action reports related to the 2002 Winter Games, an FBI major event planning conference, a USSS threat assessment and assassination seminar, and multiple cross-border critical incident and emergency response exercises. Though evidence of comparable policies across jurisdictions is not itself evidence of deliberate policy transfer (Bennett 1991: 223), the fact that the RCMP has adopted the ICS in recent years suggests that the emulation of policy instruments has also been part of this wider process of cross-border institutional learning between Canada and the USA. These inter-agency processes of learning and knowledge transfer reflect not only longstanding working relationships on matters of cross-border interest, but also raise questions about the willingness of the USA to involve itself in matters deemed to touch on its national interests.

Until recently these transnational networking activities were composed almost exclusively of state actors interacting directly with one another, but a growing range of non-state actors are now becoming prominent. The International Olympic Committee (IOC), for example, coordinates its own institutional knowledge network called the Olympic Games Knowledge Management (OGKM) program that is comparable to state institutional networks though in a non-state and transnational context. Inaugurated in 1998 as a way to reduce the growing cost and complexity of the Games, the OGKM program is geared towards "capitalizing and transferring 'know-how' from Games to Games," according to OGKM literature, by "offering a platform of learning through a variety of ways of knowledge transfer—from written reports and documents, to a global network of experienced advisors, to opportunities to learn in a live environment." As it pertains to security, this evolving "platform of learning" covers a range of issues that are the responsibility of the organizing committees such as controlling access and egress, hiring security guards and contracting police officers, acquiring security equipment, ensuring compliance with pertinent laws and regulations, and integrating venue security with the overall plans of the host law enforcement and public safety agencies.

A significant body of formal knowledge has accumulated around these responsibilities as part of the OGKM program with each organizing committee contributing after-action reports and other planning documents that are integrated by the IOC and made available to future organizing committees. These reports include detailed description of security arrangements and critical incidents as well as "a whole set of recommendations and lessons drawn from the event for how to make it smoother for future Games," according to an individual who drafted some of these reports after the 2006 Turin Games. The IOC's own "in-house" experts are also made available to organizing committees, particularly at the early stages of planning. This trade in embodied experience is supplemented by the informal practice of recruiting members of previous organizing committees—sometimes known as "Olympic nomads"—into key positions for upcoming events. VANOC's Director of Security Integration, for example, is a former Italian military official who fulfilled a comparable position in the organizing committee for the 2006 Turin Games.

International governance organizations are also becoming prominent in this field. In 2006 the European Union (EU) and the United Nations Interregional Crime and Justice Research Institute (UNICRI), which describes itself as "laboratory" for "testing ideas," embarked on a collaborative research program on major event security called EU-SEC with the aim of harmonizing national policies, encouraging practical cooperation and information sharing, and identifying and promoting best practices regarding major events across the EU. In 2006 the original EU-SEC program and its successor, EU-SEC II, were integrated as a regional platform within the UN International Permanent Observatory on Security during Major Events (IPO) along with South America and Southeast Asian platforms. The IPO functions with the aims of "strengthening international cooperation and facilitating the exchange of information among national agencies in charge of security during major events; promoting the identification of best practices within the field; improving the capability of relevant national agencies and departments to maintain security during major events" (UN 2006: 1). A range of services can be requested by member nations that support these aims, the core of which is the *Security Planning Model*. Similar to the major event templates created by the RCMP Major Events Division, this planning tool is essentially a "to-do list" for how national authorities can begin security planning for a major event. Emerging from the initial round of research conducted by EU-SEC, the planning model has since involved into the *Major Event Toolkit for Policymakers and Security Planners*, which according to an IPO official is much more detailed in that it includes specific examples and lessons from previous events. Like the previous institutional networks outlined above, the IPO has its own "in-house" experts and a roster of "hundreds" of public officials with prominent roles in previous major events, according to the same IPO official, that member states can request as consultants.

"Body shops in the brain business"

Private security firms are also becoming prominent in these institutional arrangements. Large technology companies stand out in this regard, but any private entity claiming technical expertise can seek to position itself within these networks. Insurance companies, for example, are potentially important transfer agents insofar as they may influence governments or organizing committees to adopt certain security practices as part of insurability requirements. Private companies do not, on the whole, constitute the same kind of network as those outlined above but rather shifting and temporary alliances lashed up to meet the needs of particular events. These diffuse alliances seek to insinuate themselves within state and non-state networks by monopolizing capital, or "context-specific resources" (Dupont 2004: 84), in order to increase their value. Knowledge, in this context, is capital; more specifically, practical knowledge gained through experience is capital, and promoting this capital is a vital exercise amongst private actors seeking to interact with or increase their importance within institutional networks. The slogan for the UK event management firm *Rushmans* for example is: "Knowledge is not enough—experience is essential," and promotional copy from *Rushmans* (and similar companies such as *Major Events International*) often prominently enumerates previous events in which they were involved to bolster their credentials. More importantly, private firms also deal in people, in particular former public officials with prominent roles in previous events, who can be "rented out" for long or short periods of time as consultants to governments or event organizers. Such individuals can be valuable commodities, particularly at the high end of the market, because of the capital they embody.

The US military–industrial firm *Science Applications International Corporation* (SAIC), for example, successfully procured a $350 million US technology contract from the Greek government over competing bids not only because it promoted its role as a systems integrator for the 2002 Games, but also because the company was able to recruit the former executive director of the Utah Olympic Public Safety Command (UOPSC) who was already being enticed by Greek officials to consult on the 2004 Games. This instance is consistent with SAIC's broader corporate strategy (and that of the military–industrial complex in general) to stock its ranks with former government officials valuable for their personal and professional contacts within the public sector domain from which they arrived. What is significant here is that SAIC farms out a good part of the technical work for the contracts it procures to the wider network of engineering and technology firms assembled under its corporate umbrella, which highlights how SAIC's stock-in-trade is not necessarily in the domain of technical expertise per se but in the ability to position itself within other networks by recruiting those valuable to others. As one SAIC vice-president said in an interview (himself a public safety official involved in the 2002 Games recruited to SAIC at the same time as the UOPSC executive), "one of the things that we pride ourselves in is having people with excellent experience they can bring to the table." This is why Bartlett and Steele (2007) refer to SAIC as a "body shop in the brain business."

These multinational "body shops in the brain business"—of which SAIC is only one example amongst others—are thus potentially important transfer agents by virtue of how they package and sell embodied expertise to governments and event organizers. Moreover, as security for the Games continues to move towards large-scale integration projects that only companies with military–industrial pedigree such as SAIC and similar companies can handle, these transfer agents not only conjoin different sites but also act as potential interfaces between advances in military surveillance and communications technology and their domestic application. This three-way interface of expertise is illustrated by the contingent of SAIC representatives at a 2008 major events security conference in London, that consisted of the former Salt Lake security official turned SAIC executive quoted above, a former career military official newly recruited to SAIC, and a British biometrics technology expert working on securing SAIC contracts for the 2012 London Games.

Discussion: global policy networks?

These institutional networks and private sector actors appear at first glance to constitute what Ward (2006: 70) calls "strong diffusion channels and distribution networks that exist to facilitate the transfer of policies of a particular type from one place to another." These distribution networks are composed of flows of physical artifacts—templates, models, after-action reports, and the like—that "travel in briefcases, are passed around at conferences and meetings [. . .] and [are] repeatedly the topic of discussion" (McCann 2008: 6), as well as transnational flows of people who move between events as Olympic nomads or come together regularly at the growing number of international conferences dedicated to major event security. Yet, there are numerous risks in taking evidence of these networking activities at face value as evidence of true learning. Critical takes on policy convergence warn against taking evidence of interaction as evidence of learning without detailed empirical investigation (Bennett 1991), which is often precluded in this context by veils of government secrecy. The sociology of organizational failure tells us that mistakes are prone to be repeated, particularly when lessons are drawn from experience that is distant in space and time (Donahue and Tuohy 2006).

A number of other counter-examples can be enumerated. After action reports may be drafted, circulated, revised, and passed up the chain in order to satisfy various institutional requirements, yet there is no guarantee that these reports will influence future events. "Typically, the people who are aware of these reports and who read them are either not around when it comes time to plan for the event, forget what they read, or are not in an organizational position to implement the recommendations of these reports" (Bellavita 2007: 4). It is debatable whether the globe-trotting habits of higher officials genuinely influence the activities of mid- and lower-level workers who do the bulk of the detailed work. "It depends if these people are really involved in the detailed planning," said a security planner with the organizing committee for the 2006 Games. More often than not, however, these processes occur above the heads of most: "People were just requested to work and be efficient," the same individual reported.

Public authorities and organizing committees may also hold divergent views on the look or feel of security, or in extreme cases harbor institutional animosities that hinder working relationships. A glimpse of this is gained from a confidential RCMP planning document that states: "The RCMP will provide security in spite of VANOC." Private companies may overstate their role in previous events or the value of their consultants in order to procure contracts when such involvement may have been minimal. "The world appears to be bursting with many who claim to have security experience of this Games or that," said a prominent IOC security consultant in an email communication. It is perhaps most important to recognize that processes of inter-event scrutiny are intrinsically structured by the political relations in which they are embedded. The Turin security official quoted above, for example, said in response to a question regarding the effectiveness of state-to-state observation programs, "it's debatable. [. . .] It can come down to political issues that are much bigger than us." Security for the 2004 Athens Games, for example, was influenced to a high degree, and in some cases contrary to the wishes of Greek officials, by a temporary transnational security network convened as the Olympic Security Advisory Group (OSAG) which consisted of officials from Australia, France, Germany, Israel, Spain, the UK and USA (Samatas 2011). This is an extreme example yet one that was instructive for US involvement in future Olympics (GAO 2005) and for the current analysis as it underlines how the market for best practices is not free from, but inherently structured by, prevailing relations between states.

These counterpoints should balance a tendency in official discourse and in some academic discussion of major event security to regard these processes as indicative of ongoing convergence and standardization of major event models and practices. While scrutiny of other events and the identification of lessons *do* occur, these processes are hindered by logistical factors and structured by prevailing political relations. Nonetheless, the fact that officials often *profess* to voluntarily learn from others in an ostensibly free market of ideas suggests that doing so has performative value on its own that is worth considering, such as using evidence of learning from others to justify or advance pre-formulated political aims. For example, Chinese security officials participated in many observation programs and high-level (and highly publicized) conferences in preparation for the 2008 Games (Yu, Klauser, and Chan 2009), yet Chinese security efforts overall were unmistakably domestic, suggesting that these overtures were calculated efforts to foster the perception of joining the international community rather than a genuine effort to learn from others (Thompson 2008). Statements from public officials regarding learning from others may also be seen as components of wider communicative efforts on the part of "managers of unease" (Bigo 2002) to build confidence and trust in efforts to manage what are ostensibly highly vulnerable situations. Crafted according to a political calculus weighing the persuasive benefits of standing on the shoulders of others against the political risks of *requiring* the help of others, such statements reflect the broader cultural politics of the risk society wherein authorities must not only protect but also reassure a variety of audiences that all foreseeable risks are being actively minimized (Boyle and Haggerty 2009).

One area where these networks do converge is in regard to the wider policy objectives of major event security, in particular the idea of using major events as catalysts for lasting security governance legacies. For heuristic purposes these legacies can be roughly organized into two main types: tangible outcomes such as investments in CCTV cameras, communications systems, or built infrastructure, and intangible outcomes such as improved inter-agency trust and working relationships, structural or organizational outcomes, and the development of practical expertise. While legacies of the former sort receive the majority of public and academic attention, legacies of the latter sort are of equal if not greater significance. The security legacy of the 2002 Games for the USA, for example, was less in the area of tangible outcomes and more in terms of how it served as a developmental laboratory for the structures and practices of intergovernmental cooperation and domestic crisis management that, after 9/11, would become the core of the DHS National Incident Management System. Similarly, the (still valuable) public discussion of whether the retention of CCTV cameras in Vancouver will constitute a lasting legacy for the city misses the fact that, for high-level officials involved in the 2010 Games, the development of organizational structures and practical expertise will be the most valuable legacy of the 2010 Games. This includes at one level the development of the major events template itself around the 2010 Olympics, which includes "aspects that we can duplicate for other events," according to the Director of the Major Events and Services Division, along with the development of the practical experience required to go along with such formal artifacts. "Anytime you do something big like the Olympics, it takes you through to a bigger level of experience and expertise," says Canada's federal security coordinator for the 2010 Olympics. "One has to renew these types of big events from time to time to maintain the currency of this valuable expertise" (in Frontline 2008: 38). This expertise goes beyond managing major events, however. For the head of the RCMP 2010 Integrated Security Unit the "whole of government" approach, developed and tested in the real-world laboratory of the Olympics, will be the "greatest legacy" of the 2010 Games for Canada (RCMP 2010).

The EU-SEC series of projects is explicit about promoting major events as opportunities to develop security governance expertise within the EU. A UNICRI report states: "Planners should be knowledgeable on how to ensure that countries that host major international events can gain long-term benefits from planned security. In particular, a legacy knowledge and a planning culture could ensure that the resources and know-how made available for major events such as infrastructure, training and technology solutions would enhance overall national capabilities and improve daily routine activities after the event" (UNICRI 2007: 7). The IPO, of which EU-SEC is now a regional project, has begun to promote this message beyond the European core to developing countries where major events are being held with increasing regularity yet where capabilities to manage these events may be less developed. While any country hosting an event can request assistance from the IPO, these promotional efforts are directed largely towards Central and South America and the Southeast Pacific regions where IPO Americas and IPO Asia-Pacific platforms have recently been set up. "We want nations to be able to do major event security in a way that will be useful after the event is done so that all this money invested in security is not wasted," said an IPO official in an interview. "We promote these kinds of messages," the same individual continued, which is "the only reason why there is a United Nation's program on major event security." The aim of these platforms, another IPO official explained, is primarily on fostering leadership capabilities, both structurally and individually, in these countries through mentoring services and consultation opportunities with IPO experts drawn largely from European and North American law enforcement agencies, a goal that is in keeping with the UN's wider emphasis on "capacity building" in developing countries (Ilcan and Phillips 2008).

This emphasis on expertise does not mean that technological legacies are not a part of this process. Peter Ryan, head of the New South Wales Police Department during the 2000 Sydney Games and now a prominent IOC security consultant, regularly promotes the benefit of technological outcomes for major events in conferences and other outlets such as trade magazines (see Major Events International 2009). "The preparations for the Games and the investment in security infrastructure will be an enormous legacy for the country and its national security capability after the Games are over. This opportunity should not be wasted," Ryan stated at a post-2002 Games conference (Ryan 2002: 26). This approach is particularly appealing to security companies as it directly suggests long-term government spending on security and hence new markets for their products. As such, security companies have a vested interest in shaping the discourse around major events to include security legacies. An IPO official put it this way: "To sell their products they need to convince the organizers and host governments of the major event that there are long-lasting benefits to invest in their technology and that this technology is not only for major events but can be used in other situations too." A clear example of such efforts is found with a security conference for top UK officials held in 2007 in London, devoted to "creating the legacy of 2012" as the conference was titled, convened through the sponsorship of SAIC and Northrop Grumman, both military–industrial contractors seeking large CCTV and communications integration projects being timed to coincide with the 2012 Games.

A far more attractive prospect for security companies than expanding the corporate take of the homeland security market in developed countries is for major events to be points of entry into the security markets of developing countries. Of particular interest are the so-called BRIC countries (Brazil, Russia, India, and China), where expenditures on national infrastructure are growing and the legacy potential consequentially greater. Not coincidentally, these countries will be hosting, or just have hosted, major international sporting events. American technologies firms such as IBM and Honeywell, for example, enthusiastically exploited the confluence of China's virtually untapped security market with the government's wider policy objectives to combine the $6.5 billion (US) "Grand Beijing Safeguard Sphere" along with preparations for the 2008 Games to sell a range of technologies to the authoritarian government (Bradsher 2007; SIA 2007).

This profit-driven search for new and expanding markets merges neatly with the IPO's newest initiative—IPO TECH—which aims to develop partnerships with "prominent technology suppliers," to provide "proven" security technologies to the organizers of major events and to train practitioners in its use. Given that this initiative would provide a direct pathway into coveted markets it is little wonder that numerous multinational security companies are seeking to establish themselves as partners of this IPO initiative.

Conclusion

This chapter has provided insights into security governance at major events by focusing on how different knowledge networks come together around these security issues and has considered their overall impact. This analysis has focused on three main institutional networks, each of which have the explicit purpose of facilitating the movement of knowledge within an institution, along with the role of private industry as brokers of embodied expertise. While high-level representatives from these networks and private industry may travel the world to observe other events, meet regularly at international government and industry conferences, host debriefing sessions to examine "what works," and publicly profess the importance of learning from the past, it cannot be directly inferred that these activities amount to the disinterested free market of best practices they are often presented to be. The informal realities of bureaucratic and institutional structures hinder the uptake of ideas and best practices, which are further structured by prevailing power relations between states, pre-formulated political aims, and the ongoing search for new markets. These points support Peck and Theodore's argument that "policy models that affirm and extend dominant paradigms, and which consolidate powerful interests, are more likely to travel with the following wind of hegemonic compatibility or imprimatur status" (Peck and Theodore 2010: 170).

The idea of proactively positioning major events as catalysts for security legacies travels easily as it meshes with a number of dominant interests. For state agencies it represents an opportunity to maximize a return on what is often a substantial investment in security for one-time events. Of course, this desire can tip into ill-conceived buying sprees, deliberately padded budgets, and in some cases graft. The IOC has an implicit interest in promoting this approach—or at least not discouraging it—so as to identify an upside for governments and the public for the billion dollar budgets Olympic security now commands. When visiting Vancouver before the 2010 Games, a visit that coincided with a rash of murders related to the city's drug trade, Jacques Rogge responded to questions about Canada's $1 billion CAN security budget by assuring the reporter that it would help fight crime in Canada "for decades to come" (Mickleburgh 2009). Security legacies fit the UN's mandate for capacity building in developing countries, in this case by using major events as "teachable moments" in fostering security subjectivities and structures. For the security industry it provides a powerful vehicle for expanding into new markets and revenue streams. The confluence of these interests makes the Games powerful sites in the production and globalization of security expertise in the post-9/11 world, particularly in relation to cities of the global south where major events will be staged with increasing regularity.

At least two normative questions are raised by the consolidation of this new "common sense" approach to major events. First, one needs to examine how substantial investments in security and surveillance capabilities exacerbate levels of social polarization in cities that are already deeply polarized, such as in Rio de Janeiro where there is already talk of rapidly modernizing the city's police and security infrastructure around the 2014 FIFA World Cup and 2016 Olympics. Closely related to this is whether the idea of major event security legacies provides justification for authoritarian governments to increase control over its citizens, such as appears to be the case in China during and after the 2008 Games, something that occurred in full view of the world and with the assistance of western multinationals. These questions are particularly pressing given that major events are becoming coveted moments for cities and nations of the global south as a path to first-class global citizenship (Bernhard and Martin 2011), so it is of prime importance to examine the wider socio-political effects of major events and their security legacies for urban inhabitants.

By way of conclusion it is worth briefly considering something that has been neglected in this chapter. It follows from the above points that there is also a politics of knowledge *im*mobilities (Adey 2006) around major events for those ideas that do not mesh with the prevailing contours of the major events industry. Counter-movements, for one, face brisk headwinds in criticizing the power and politics of these global spectacles. Yet, as much as major events offer opportunities for authorities to learn from one another and foster professional networks they also offer comparable opportunities for counter-movements of various kinds to galvanize their own networks. In the year before the 2010 Games, Vancouver played host to a steady stream of activists, artists, academics, and others questioning all manner of issues related to the Games, culminating in February 2010 in some of the largest protests Vancouver and Canada has seen in recent years. While the policing of these protests may have provided lessons for colleagues near and far, so too do these resistance efforts offer valuable lessons for counter-movements around the world which have become part of an ongoing legacy of critique and resistance to the status quo.

References

Adey, P. 2006. "If mobility is everything then it is nothing: Towards a relational politics of (Im)mobilities." *Mobilities* 1 (1): 75–94.

Bartlett, D. and J. Steele. 2007. "Washington's $8 billion Shadow." *Vanity Fair* (online). URL: http://www.vanityfair.com/politics/features/2007/03/spyagency200703? (accessed September 24 2008).

Bellavita, C. 2007. "Changing homeland security: A strategic logic of special event security." *Homeland Security Affairs* 3 (3): 1–23.

Bennett, C. 1991. "What is policy convergence and what causes it?" *British Journal of Political Science* 21 (2): 215–233.

Bernhard, D. and A. Martin. 2011. "Rethinking security at the Olympics." In *Security Games*, edited by Colin Bennett and Kevin D. Haggerty, London: Routledge.

Bigo, D. 2002. "Security and immigration: Towards a critique of the governmentality of unease." *Alternatives* 27 (1): 63–92.

Boyle, P. and K. Haggerty. 2009. "Spectacular security: Mega-Events and the security complex." *International Political Sociology* 3 (3): 257–274.

Bradsher, K. 2007. "September 11. An opportunity for Wall Street in China's surveillance boom." *New York Times* A.1.

Dolowitz, D. and D. Marsh. 1996. "Who learns what from whom: A review of the policy transfer literature." *Political Studies* 44 (2): 343–357.

Donahue, A. and R. Tuohy. 2006. "Lessons we don't learn: A study of the lessons of disasters, why we repeat them, and how we can learn them." *Homeland Security Affairs* 2 (2): 1–28.

Dupont, B. 2004. "Security in the age of networks." *Policing and Society* 14 (1): 76–91.

Frontline. 2008. "Federal security support to the 2010 Olympics: An interview with Ward Elock." *Frontline Security* Spring: 36–38.

GAO. 2005. *Olympic Security: U.S. Support to Athens Games Provides Lessons for Future Olympics.* Washington, D.C.: United States Government Accountability Office.

Haas, P. 1992. "Epistemic communities and international policy coordination." *International Organization* 46 (1): 1–35.

Ilcan, S. and L. Phillips. 2008. "Governing though global networks: Knowledge mobilities and participatory development." *Current Sociology* 56: 711–734.

Johnston, L. and C. Shearing. 2003. *Governing Security: Explorations in Policing and Justice.* New York: Routledge.

Klauser, F. 2009. "Interacting forms of expertise in security governance: The example of CCTV surveillance at Geneva's international airport." *The British Journal of Sociology* 60 (2): 279–297.

Major Events International. 2009. "Q&A: An interview with Dr. Peter Ryan." *Major Events International* (online). URL: www.majoreventsint.com (accessed September 9, 2009).

McCann, E. 2008. "Expertise, truth, and urban policy mobilities: global circuits of knowledge in the development of Vancouver, Canada's 'four pillar' drug strategy." *Environment and Planning A* 40 (4): 885–904.

Mickleburgh, R. 2009. "Olympic security efforts to help curb crime in Canada, IOC leaders says." *The Globe and Mail* February 12, A.9.

Peck, J. and N. Theodore. 2010. "Mobilizing policy: models, methods, and mutations." *Geoforum* 41 (2): 169–174.

RCMP. 2005. *Planning Update.* Vancouver: Vancouver 2010 Integrated Security Unit.

RCMP. 2010. A/Commr Bud Mercer on the challenges of securing the Olympic Games. *Vancouver 2010 Integrated Security Unit* (online). URL: www.v2010isu.ca (accessed March 12, 2010).

Reese, S. 2008. *National Special Security Events.* Washington, D.C.: Congressional Research Service.

Rose, R. 1991. "What is lesson drawing?" *Journal of Public Policy* 11 (1): 3–30.

Ryan, P. 2002. *Olympic security: the relevance to homeland security.* Salt Lake City: The Oquirrh Institute.

Samatas, M. 2011. "Surveilling the 2004 Athens Olympics in the aftermath of 9/11: International pressures and domestic implications." In *Security Games*, edited by Colin Bennett and Kevin D. Haggerty. London: Routledge.

SIA. 2007. *China Security Market Report Special Supplement: Olympic Update.* Alexandria, Virginia: Security Industry Association.

Stone, D. 2004. "Transfer agents and global networks in the 'transnationalization' of policy." *Journal of European Public Policy* 11 (3): 545–566.

Thompson, D. 2008. "Olympic security collaboration." *China Security Review* 4 (2): 46–58.

Thompson, L. 2009. "Major event marathon: Preparing for the main event, or in this case main events." *The Frontline Perspective* 3 (3): 1–3.

UN. 2006. *Background Note: International Permanent Observatory on Security Measures during Major Events.* New York: United Nations.

UNICRI. 2007. *Towards a European House of Security at Major Events.* Turin: United Nations Interregional Crime and Justice Research Institute.

Waddington, D. 2008. *Public Order Policing: Theory and Practice.* Portland, OR: Willan Publishing.

Ward, K. 2006. " 'Policies in motion,' urban management and state restructuring: the trans-local expansion of business improvement districts." *International Journal of Urban and Regional Research* 30 (1): 54–75.

Yu, Y., F. Klauser, and G. Chan. 2009. "Governing security at the 2008 Beijing Olympics." *International Journal of the History of Sport* 26 (3): 390–405

Index

accountability 156
accumulation *see* capital accumulation
activists viii, 6, 38, 39, 65, 77, 157, 182
   see also protest, 'renditions', terrorism
advertising/transnational sponsorships 7, 44, 87–98, 150
   see also capital accumulation, consumption/commercialization, securitization: connections to commercialization
agency *see* autonomy of local actors
Aichi Expo 2005 *see* EXPO World's Fair
airspace
   fighter jets 126, 133, 160
   surveillance 1, 94, 126
   see also air travel (commercial), blimps, militarization, missile defense, satellite surveillance, unmanned aerial vehicles
air travel (commercial) 24–5, 28, 38
Albertville 1992 *see* Olympic Games
American Technology Corporation 160
   see also long range acoustic device, private security/surveillance contractors
Athens 2004 *see* Olympic Games
Atlanta 1996 *see* Olympic Games
Austria/Switzerland 2008 *see* European Football Championships
authoritarianism 1, 41, 78, 148, 182
autonomy of local actors 17, 121, 132–3, 120–33
   UEFA versus Olympic Games 133
   see also governance, sovereignty

Barcelona 1992 *see* Olympic Games
Beijing 2008 *see* Olympic Games
bid requirements 28, 37, 50, 56, 66, 127, 138

biometrics 1, 3, 8, 12, 40, 45, 47, 78, 80, 81, 96, 124, 133, 160, 177
   see also databases, data retention, face recognition, fingerprinting, ID cards, identification technology, voice analyzers
blimps 12, 25, 62, 66, 163
   see also airspace surveillance
bomb/explosives detection 93, 144
boosterism 88, 125
   see also bribery/corruption, private security/surveillance contractors, mega-events: private sector influence on, transnational corporations
borders viii, 1, 4, 38, 42, 81, 93, 111, 112, 122, 124, 147, 155, 174
branding
   regulations 89–91, 98
   of events 7, 16, 48, 94, 98, 146, 157
   of locations *see* destination marketing
   of nations 97, 109
      see also international relations, legacies: for developing nations, mega-events: hosting as a marker of status, performance
   of products *see* advertising/transnational sponsorships, consumption/ commercialization, food monopolies, mega-events: private sector influence on
   of security technologies 180
      see also mega-events: hosting as a marker of status, private security contractors, spectacular security
bribery/corruption 7, 55–6, 60–1, 66, 108, 181
'broken-windows theory' 109
   see also crime

Index

Calgary 1988 *see* Olympic Games
capital accumulation 150–64, 176
   securitization: connections to commercialization, transnational corporations, neoliberalism, private security/surveillance contractors
capitalism 77, 150–64
   see also capital accumulation, transnational corporations, neoliberalism, private security/surveillance contractors
   CCTV 8–9, 25, 28, 40, 42, 43–5, 47, 49, 50, 56, 60–2, 66, 79, 80, 89, 93, 95–8, 104, 113, 124, 127, 128, 131–3, 140, 142, 143, 145, 148, 160, 163, 179, 180
   see also cyclops, face recognition, intelligent passenger service program
chimera incidents 114–15
civil liberties 2, 4, 6, 13–14, 16, 55–6, 60, 66, 67, 94, 105, 139, 148, 154, 156–7
   see also democracy, free speech, protest, social welfare
claims makers 12, 23, 61, 111, 116, 159
class conflict 152
cloud computing 4
command centres *see* operation/command/control centres
commercialization *see* consumption/commercialization
communication technology 16, 93, 126, 129–31, 146, 147, 177, 179
   see also information technology, wiretapping
consumption/commercialization 17, 51, 77, 87–98, 109–10, 121, 123, 125, 150–64
   see also advertising/transnational sponsorships, capital accumulation, food monopolies, securitization: connections to commercialization, transnational corporations
Contemporary Security Canada (CSC) 160
   see also private security/surveillance contractors
control centres *see* operation/command/control centres
control societies 104–5, 115
corruption *see* bribery/corruption
counter-terrorism exercises 31, 58, 59, 127, 143

crime 1, 9, 11, 26, 42, 45, 55, 77, 80, 83, 93, 96, 109–10, 115, 181
   see also broken-windows theory, legacies: increasing local criminal activity, marginalization: ban on begging, organized crime, signage by-laws, protest: policing dissent, hooliganism: criminalization of football fans
crisis response/management 58, 131, 138, 148, 162, 172, 174, 179
crowd control *see* long range acoustic device, protest, security cordons/rings
cultural conformity 83, 84
   see also deviance, othering, self-regulation
cyclops video-surveillance system 8, 131–2
   see also CCTV
C4I security system 9, 25, 27–8, 56, 60–3, 160, 163
   see also SAIC, Siemens, 'dummy project'
C4ISR security system 94

databases 9–10, 13, 28, 62, 93, 94, 138, 163
   see also 'data doubles', data exchange, data retention, dataveillance, *Gewalter Sport*, legislation: data protection, watch lists
'data doubles' 3
   see also databases
data exchange 110, 111, 112, 126, 140, 146, 153, 155
   see also knowledge: construction and transfer
data protection authorities, limited power of 14
   see also legislation: data protection
data retention 3, 9–11, 155
   see also databases, data protection authorities, legislation: data protection
'dataveillance' 25, 62, 139
democracy viii, 2, 7, 14, 66, 77, 80, 91, 95, 105, 111, 116, 157
   violations of constitutional principles 10, 64, 66, 92, 93, 95, 97, 112, 156
   see also civil liberties, free speech, social welfare
de-securitization 23
   see also securitization

destination marketing (tourism) vii, 5–6, 20, 88, 150, 154–5
  via place branding 6–7, 41, 88, 106, 107, 108, 113, 138, 151, 161
  see also branding, consumption/commercialization, risk: perception of inhibiting tourism
developing nations *see* legacies: for developing nations
deviance 92, 93
  see also cultural conformity, self-regulation
disciplinary societies 104–5
  see also enclosure, panopticon
discrimination 5, 76, 111, 115
  see also marginalization, othering
disease surveillance *see* health surveillance
displacement *see* gentrification, marginalization: relocation of vulnerable individuals, spatial displacement
'dispositif' 121
drones *see* unmanned aerial vehicles
drug testing 10
  see also World Anti-Doping Agency
'dummy project' 61
  see also C41, SAIC, Siemens, Athens 2004

economic impact *see* legacies: economic impact
emergency
  permanent state of 32, 104
  see also crisis response/management, knowledge: construction and transfer of, risk, security: threats as socially constructed
enclosure 104
  see also disciplinary societies
environmental impact *see* legacies: environmental impact
EU *see* European Union
European Football Championships (UEFA) 107, 121–2, 126, 128–9, 132
  Portugal 2004 146
  Austria/Switzerland 2008 16–7, 92, 98, 120–33
European Union (EU) 17, 60, 120, 122, 127, 175
exceptionalism 20, 31, 42, 46, 129–32, 151
  see also spaces of exception, state of exception

exclusion *see* marginalization
experimentation vii, 3, 11, 56, 60, 98, 120, 163, 179
expertise 8, 17, 27, 120–1, 125, 127, 169–82
  as business opportunity 125, 162, 163, 180, 182
  see also knowledge: construction and transfer of, private security/surveillance contractors
EXPO World's Fair 103
  Osaka Expo 1970 74
  Aichi Expo 2005 75
  Shanghai Expo 2010 42

face recognition 8, 12, 45, 80, 96
  see also biometrics, CCTV, fingerprinting, voice analyzers
fan zones/miles/parks 11, 16, 89–90, 92, 95, 97, 112–13, 123, 126, 127
  see also consumption/commercialization, security cordons/rings, securitization: connections to commercialization
fear 23, 27, 79
  climate of viii, 22, 24, 64, 74–7, 111, 142, 161
  see also knowledge: construction and transfer of, media representations, risk
festivalisation *see* spectacle, urban development: festivalisation
FIFA: *see* International Federation of Association Football
  see also FIFA World Cup
FIFA World Cup viii
  France 1998 107
  Japan/Korea 2002 16, 72–84
  Germany 2006, 6, 9, 11, 13, 14, 16, 49, 87–98, 103–16, 126, 127
  South Africa 2010 1, 98, 109, 115–16
  Rio de Janeiro 2014 182
fingerprinting 1, 78, 124, 133, 160
  see also biometrics, face recognition, voice analyzers
Finmeccaica (Elsag Datamat) 141
  see also integrated security system
food monopolies 90–1, 93, 97, 156
  see also advertising/transnational sponsorships, securitization: connections to commercialization, transnational corporations
France 1998 *see* FIFA World Cup

//

Index

free speech 156, 157
   see also civil liberties, democracy, protest, signage by-laws
function creep 8, 9, 49
   see also legacies: informational, legacies: technological

Garret Metal Detectors 141, 160
   'illusory security' 160
   see also private security/surveillance contractors
gentrification 49, 50, 89, 110, 153–4
   see also urban development, marginalization: relocation of vulnerable individuals
Geospatial Information System 147
Germany 2006 see FIFA World Cup
Gewalter Sport (German data bank of violent sports offenders) 9, 92, 94
globalization 4, 73, 75, 81, 83, 121, 126, 137, 147, 171, 182
   see also transnational corporations, international relations, sovereignty
goodwill, feelings of 37, 48, 74
   see also public morale, public opinion, social cohesion
governance 108, 114, 120, 151–3, 161, 163, 179
   self-regulation 103–16, 115
   state 72–84, 92, 93, 95, 182
     see also social control, sovereignty
   via private interests see private security/surveillance contractors, transnational corporations
   via a transnational ruling class 84, 92, 98, 104–5
     see also advertising/transnational sponsorships, consumption/commercialization, neoliberalism, private security/surveillance contractors, transnational corporations
   see also nodal governance, surveillant governmentality

health risks 139, 160
health surveillance 9–10
Hokkaido 1972 see Olympic Games
Honeywell 159, 160, 163, 180
   see also Perimiter Intrusion Detection System, private security/surveillance contractors, SAIC

hooliganism 8, 76, 79, 80, 82, 83, 92, 94, 108, 110, 111, 112, 115, 126, 128
   'Hooligan-Knast' 113
   criminalization of football fans 113
   see also Gewalter Sport

IBM 158, 159, 180
   see also private security/surveillance contractors
ID cards 78, 81
   see also biometrics, identification technology
identification technology 84, 93
   see also biometrics, databases, ID cards, radio-frequency identification devices
inequality see marginalization
information technology 3, 11, 93, 147
   see also cloud computing, communication technology, databases, data retention, key-word intercepts
insurance companies 175
   see also claims makers, knowledge: construction and transfer of, mega-events: private sector influence on, risk
Integrated Security System 7, 137–48
   see also Finmeccaica
intelligence agencies 10, 22, 24, 26, 31, 38, 57, 58–9, 61, 64, 78, 79, 93, 94, 143, 146–7, 153, 157, 158, 170, 172
   see also key-word intercepts, knowledge: construction and transfer of, 'renditions'
Intelligent Passenger Service Program (IPS) 45
   see also CCTV, transit
International Federation of Association Football (FIFA) vii, 2, 7, 87–98, 103–16, 126, 132
   see also FIFA World Cup
International Olympic Committee (IOC) vii, 2, 7, 10, 16, 27, 28, 32, 36, 38, 41, 42, 46, 48, 50, 55, 89, 126, 132, 140, 141, 143, 157, 162, 174–5, 180, 181
international relations 16, 31, 32, 55, 60, 64, 72, 73–7, 112, 133, 137, 138, 159, 173–4, 178, 179–80, 181, 182
   see also globalization, legacies: economic impacts of, legacies: for developing nations, mega-events: hosting as a marker of status, performance, 'security spectacle'

interoperability 13, 162–3
  see also standardization
IOC see International Olympic Committee

Japan/Korea 2001 see Olympic Games
Japan/Korea 2002 see FIFA World Cup

key-word intercepts 59
  see also information technology, intelligence agencies
kidnapping 59
  see also intelligence agencies
knowledge
  assemblages 171–7
  construction and transfer of 39, 59–60, 92, 93, 98, 105, 111, 114 (Table 6.2), 116, 120, 126, 127, 128, 140, 146–7, 162–3, 169–82
  'Olympic gypsies' 175, 177
  see also data exchange, expertise, risk, security: threats as socially constructed, standardization

Lake Placid 1980 see Olympic Games
legacies of mega-events 7–12, 17, 37, 48–9, 50, 55, 60, 67, 96–7, 108, 132, 151
  cultural acceptance/normalization of surveillance 12, 113–14, 150
    see also surveillance: intensification of
  developing nations, for 7, 20, 29, 32, 179–80, 181, 182
  economic impact 14, 15, 16, 21, 25, 28, 29, 51, 55–67, 92, 97, 104, 106–9, 123, 140–2, 153, 154, 158, 163
    see also tax exemptions
  environmental impact 14, 51, 139, 140–2, 148, 157, 163
  human health see health risks
  increasing local criminal activity 50–1
  informational see databases, data retention, function creep
  international relations see international relations
  legal 11, 90, 93, 98, 105, 110, 111, 112–13, 116, 128, 131–2, 142, 146, 155, 161
  reconfigured relationships between public/private agencies 161–4, 169, 179, 181
    see also governance: transnational ruling class, neoliberalism
  social welfare see social welfare, marginalization
  spatial or infrastructural 11–12, 17, 49, 51, 89, 95, 103–16, 128, 133, 137–40, 179
    see also urban development, gentrification, transit, marginalization: relocation of vulnerable groups
  technological 8–9, 17, 37, 43, 49, 55, 79, 83, 96–8, 128–9, 129–32, 133, 139, 161–2, 179, 180
    see also function creep, surveillance: intensification of
legislation 9, 14, 49, 64, 88, 97, 105, 122, 130–2, 133, 137, 146, 148, 155, 156, 164, 170, 173
  data protection 8, 9, 11, 14, 96, 97, 131
  enabling mega-events see legacies, legal
  circumvention of 9, 10, 64, 66, 92, 97, 160
  convergence of 42, 128, 170, 175, 177
  privacy 5, 8, 14
  violations of see democracy: violation of constitutional principles
    see also signage by-laws, marginalization: ban on begging
Lillehammer 1994 see Olympic Games
London 2012 see Olympic Games
Long Range Acoustic Device (LRAD) 160
  see also American Technology Corporation, health risks, protest
Los Angeles 1984 see Olympic Games

marginalization 11, 17, 51, 78, 105, 108–9, 151–2, 157
  ban on begging 113
  fear of 105
    see also self-regulation
    see also gentrification
  policing/criminalizing the poor 6, 44–5, 93, 152, 154–6, 157, 164
  relocation of vulnerable individuals 51, 89, 92–93, 104, 109, 116, 153–5
    see also discrimination, social sorting, social welfare
marketing see branding, destination marketing, consumption/commercialization
media representations 2, 9, 12, 16, 21, 22, 29, 38, 41, 55, 56–8, 75, 76, 79, 80, 81, 89, 98, 103–16, 107, 111, 113, 114 (Table 6.2), 140, 142, 157, 169, 181

mega-events
    cost/benefit analysis of hosting 108
        (Table 6.1), 116
    hosting as a marker of status 6, 15,
        20–32, 67, 72–3, 75, 88, 107–9,
        120, 138, 140, 154, 159, 161–2, 182
    impact on nation/local community
        see legacies
    and International relations see
        International relations
    private sector influence on 13, 56, 59,
        60–3, 66, 87–98, 103, 110, 120,
        129–33, 158–63, 175–7
        see also advertising/transnational
            sponsorships, insurance industry,
            private security/surveillance
            contractors
    resistance to see protest
    state structure and impact on 13, 120–33
Melbourne 1956 see Olympic Games
Mexico 1968 see Olympic Games
militarization 26–8, 30, 41, 42, 59, 61, 93,
    98, 112, 121, 123, 125–6, 139, 141,
    144–5, 153, 158–9, 160, 162–3,
    176–7
    of urban space 125
    see also airspace surveillance, military-
        industrial complex, missile defense,
        unmanned aerial vehicles, warfare
military-industrial complex 176
    see also militarization
missile defence 25, 27
    see also militarization, nuclear
        proliferation, warfare
mobile terrestrial trunked radios
    (TETRAs) 61, 147
mobility
    restrictions on individual 94, 112,
        143–4
        see also air travel, borders,
            hooliganism, marginalization:
            relocation of vulnerable
            individuals, 'renditions', security
            cordons/rings, transit, watch lists
    tracking of individual see radio-
        frequency identification devices,
        vehicle tracking
Montreal 1976 see Olympic Games
Moscow 1980 see Olympic Games
Munich 1972 see Olympic Games

national identity 16, 82, 97, 108
NATO 28, 59, 60, 94, 112, 126

neocommunitarianism 87–8, 97, 98
neoliberalism 12, 16, 72, 83–4, 87–98,
    109, 10, 150–64, 162, 169
    definition of 91
    see also capital accumulation, legacies:
        reconfigured relationships between
        public/private agencies, mega-
        events: private sector influence on,
        private security/surveillance
        contractors, transnational
        corporations
nexus approach 151, 163
'nodal governance' 47
    see also governance, nodal security,
        operation/command/control
        centres
'nodal security' 47
    see also nodal governance, operation/
        command/control centres
normalization see legacies: cultural
    acceptance/normalization of
    surveillance, public opinion,
    surveillance: intensification of
Nuclear proliferation 29–31
    see also missile defence, warfare

'Olympic difference' 15, 20–1, 26
Olympic Games
    changes to following 9/11 5, 16, 55–67,
        158, 179, 182
        see also terrorist acts: 9/11
    1956 Melbourne 76
    1964 Tokyo 16, 41, 49, 73, 74, 76
    1968 Mexico 39
    1970 Sapporo 73–4, 75, 76, 77, 78
    1972 Hokkaido 76
    1972 Munich 36–7, 40, 107
    1976 Montreal 40–1
    1980 Lake Placid 40, 41, 142
    1980 Moscow 40, 41
    1984 Los Angeles 41
    1984 Sarajevo 36, 142
    1988 Calgary 37
    1988 Seoul 37–8, 41, 49, 154
    1992 Albertville 37, 43
    1992 Barcelona 37, 38, 41, 107, 154
    1994 Lillehammer 37
    1996 Atlanta 37, 40, 43, 154, 155
    2000 Sydney 11, 37, 40, 41, 49, 61, 63,
        138, 155, 156, 158, 162, 180
    2001 Japan/Korea 146
    2002 Salt Lake City 61, 65, 138, 146,
        158, 159, 173–4, 176, 179, 180

Index   191

2004 Athens vii, 8, 9, 13, 16, 24, 25, 27–8, 31, 37, 41, 55–67, 97, 107, 120, 126, 127, 137, 138, 146, 148, 158–60, 163, 176, 178
2006 Turin 9, 11, 17, 126, 137–48, 158, 173, 175, 178
2008 Beijing 2, 25, 27, 28, 31, 40, 41–2, 63, 154, 158–9, 160, 178, 180, 182
2010 Vancouver/Whistler vii, 6, 9, 11, 14, 17, 24–5, 27, 28, 92, 98, 150–64, 169, 171, 173, 181, 182
2012 London 6, 7, 11–12, 13, 14, 16, 36–51, 158, 159, 177, 180
2016 Rio de Janeiro 29, 42, 182
'Olympic gypsies' *see* knowledge: Olympic gypsies, expertise
operation/command/control centres 25, 47, 59, 62, 142, 143, 145, 160, 163
see also militarization
organized crime 11, 46, 50, 126
see also crime, legacies: increasing local criminal activity
Orwell, George 3
Osaka Expo 1970 *see* EXPO World's Fair
'othering' 73–6, 81–4
see also discrimination, marginalization

panhandling *see* marginalization
panopticon 3, 60, 104
see also disciplinary societies, self-regulation, super-panopticon
performance 17, 21, 31, 72, 74, 108, 116, 138, 148, 159, 161–2, 178, 182
see also branding, international relations, legacies: developing nations, mega-events: hosting as a marker of status, 'security spectacle'
Perimeter Intrusion Detection System (PIDS) 160
see also Honeywell, SAIC
polycom 129–31, 132, 133
see also radio communication, Siemens
Portugal 2004 *see* European Football Championships
poverty 84, 151, 154
see also legacies: economic impact, marginalization: policing/criminalizing the poor, social welfare
pre-emptive risk assessment 5, 26, 124, 147, 153, 157
see also expertise, risk

prestige *see* mega-events: hosting as a marker of status
privacy 13, 66, 132, 148
as an inadequate critique 5
legislation *see* legislation: privacy, legislation: data protection
private security/surveillance contractors 46, 47, 49, 56, 60–3, 66, 67, 91, 93, 94–5, 97, 104, 105, 110, 120, 125, 126, 127, 129–32, 152–3, 158–63, 164, 169–70, 171, 175–81
see also capital accumulation, legacies: reconfigured relationships between public/private agencies, mega-events: private sector influence on, military-industrial complex, neoliberalism, 'Olympic gypsies', security-industrial complex, surveillance-industrial complex, surveillance: intensification of
privatization *see* neoliberalism
profit *see* capital accumulation, consumption/commercialization, private security/surveillance contractors, transnational corporations
protest 6, 14, 39, 48, 81–2, 98, 108–9, 113, 115, 116, 138–9, 140–2, 147, 148, 154–5, 157, 182
policing dissent 154, 156–8, 163–4, 182
see also activists, free speech, long range acoustic device, 'renditions', public opinion, security: threats as socially constructed, student movements, signage by-laws, terrorism, violence
public opinion vii, 13, 15, 16, 17, 22, 23, 26, 40, 48, 58, 64, 66, 67, 80, 107, 109, 113, 115, 132, 133, 137–9, 142, 148, 151, 154, 156
see also legacies, media representations, protest, surveillance: legitimation of, trust
public morale 67, 106, 108
see also goodwill: feelings of, social cohesion

RA *see* regulation approach
racism *see* discrimination
radio communication 125, 128–31, 147
see also mobile terrestrial trunked radios, polycom

radio-frequency identification devices (RFID) 1, 3, 12, 28, 41, 47, 48, 93, 96, 124, 128, 139
regulation approach (RA) 151–2
  see also neoliberalism, capitalism, governance
'renditions' see kidnapping
resistance see protest
RFIDs: see radio-frequency identification devices
Rio de Janeiro 2014 see FIFA World Cup
Rio de Janeiro 2016 see Olympic Games
risk 7, 10–11, 16, 17, 20, 22–6, 31, 37, 39, 43, 50, 66, 74, 76, 91–2, 105, 111–12, 126, 138, 148, 152, 158, 169, 178
  perception of inhibiting tourism 111, 161
  see also expertise, knowledge: construction and transfer of, 'Olympic gypsies', pre-emptive risk assessment, security: threats as socially constructed

SAIC (Science Applications International Corporation) 7, 56, 58, 60–3, 159, 160, 163, 176–7, 180
  see also C41, 'dummy project', military-industrial complex, Perimeter Intrusion Detection System, private security/surveillance contractors, Siemens
Salt Lake City 2002 see Olympic Games
Sapporo 1970 see Olympic Games
Sarajevo 1984 see Olympic Games
satellite surveillance 1
  see also airspace surveillance
securitization 4, 5, 16, 22, 23, 88, 105, 120, 137, 150
  connections to commercialization 89–91, 98, 150–64
  see also de-securitization
security
  assemblage 16, 88
  as symbolic investment 2, 20–32, 88, 109, 159
  contractors see private security/surveillance contractors
  cordons/rings 11, 23, 27, 28, 42, 44, 46–8, 49, 51, 90–1, 93, 95, 123, 143–4
    see also fan zones/miles/parks, spaces of exception
  definition of 22–3
  following 9/11 see terrorist attacks: 9/11

industrial complex vii, 7, 13, 21, 55, 153, 158, 162, 164
  see also private security/surveillance contractors
industry associations, requiring further empirical investigation 159
threats as socially constructed 4, 6, 10, 16, 17, 22–7, 37, 50, 66, 78–82, 92, 105, 111, 114 (Table 6.2), 148, 155–9
  see also risk, knowledge construction and transfer of
spectacle 2, 67, 87, 137, 150
self-regulation see governance: self-regulation
Seoul 1988 see Olympic Games
Shanghai Expo 2010 see EXPO World's Fair
Siemens 7, 56, 58, 60–3, 96, 127, 130, 160, 163
  see also C41, 'dummy project', private security/surveillance industrial contractors, SAIC
signage by-laws 6, 11, 156
  see also free speech, protest
snipers 93
social cohesion 87–8, 92, 97, 108, 141
  see also goodwill: feelings of, national identity, public morale
social control 6–7, 11, 14, 40, 74, 83, 92, 98, 105, 108, 113, 151–2, 182
  see also governance
social sorting 5, 51, 82, 92
  see also discrimination, marginalization
social welfare 97, 150, 163
  see also legacies: economic impacts, poverty, marginalization, health risks
South Africa 2010 see FIFA World Cup
sovereignty
  local stakeholders 121
  organizational 89, 93, 98
  state 13, 16, 55–6, 59, 60, 63–4, 66–7, 121, 150
  see also agency, autonomy, governance
'spaces of exception' 11, 36, 39, 42, 43–4, 49, 82, 157
  see also exceptionalism, security cordons/rings, state of exception
spatial control 112–15
'spatial displacement' 2, 37, 48
  see also terrorism
spectacle 12, 17, 26, 65, 87, 150, 151
  see also performance, security: spectacle

sponsors *see* advertising/transnational sponsorships
'spychips' *see* radio-frequency identification devices
standardization viii, 15, 36–51, 92, 105, 111, 112, 113, 116, 120–1, 125, 127–9, 132, 172, 178
see also interoperability
state of exception 12, 32, 113
see also exceptionalism, spaces of exception
status *see* mega-events: hosting as a marker of status
'strategic incapacitation' 93
student movements 78
see also protest
suasive surveillance 4, 83
summits 2, 73, 74, 75, 81, 89, 122–3, 126, 156, 160, 162, 173
super-panopticon 16, 55, 56, 61–2
see also panopticon
surveillance
 contractors *see* private security/surveillance contractors
 definition of 3, 4
 Government protection of commercial interests via 156
  see also consumption/commercialization, fan zones/miles/parks, securitization: connections to commercialization
 industrial complex 21, 55, 60, 164
  see also private security/surveillance contractors
 legitimation of 12, 16, 49, 105, 110, 113, 132, 150, 156, 157
 public reaction to *see* public opinion
 society 5, 83,
 the intensification of 2–3, 5, 41, 42, 44–5, 93, 96–7, 104, 108, 110, 139, 148, 156, 158–60, 164
 see also transparency: new era of
surveillant assemblage 3, 47, 88, 95, 133
surveillant governmentality 72
see also governance, neoliberalism
Sydney 2000 *see* Olympic Games

tax exemptions 97
see also capital accumulation
technologicalization 121, 124
technological legacies *see* legacies: technological
teleological reasoning 114–15

terrorism 2, 6, 11, 16, 21, 24, 25, 36–8, 43, 46, 50, 57, 59, 77–8, 80, 81–2, 110, 112, 126
 the Olympics as a symbolic target for 38, 46
 see also spatial displacement, terrorist attacks, war on terror
terrorist attacks 42, 110, 115, 127
 Barcelona 38
 London 11, 43, 44, 46, 81, 137, 142
 Madrid 38
 Munich 1974 36–7, 39, 78
 Seoul 38
 Tokyo 75
 9/11 4, 7, 16, 21, 24, 36–8, 41–2, 55–8, 64, 66, 67, 79, 81, 105, 110, 171
 see also terrorism, war on terror
TETRAs *see* mobile terrestrial trunked radios
Tokyo 1964 *see* Olympic Games
tourism *see* branding, consumption/commercialization, destination marketing, risk: perception of inhibiting tourism
transit 17, 96–7, 107, 110, 124, 138–40, 142, 146, 148, 150, 153
 see also intelligent passenger service program, legacies: spatial and infrastructural, vehicle tracking, mobility
transnational corporations *see* advertising/transnational sponsorships, capital accumulation, consumption/commercialization, governance: transnational ruling class, insurance industry, mega-events: private sector influence on, food monopolies, private security/surveillance contractors, securitization: connections to commercialization
transparency
 new era of 4
trust 13, 14, 24, 57, 64–5, 142, 178, 179
 see also public opinion, international relations
Turin 2006 *see* Olympic Games

UAE *see* unmanned aerial vehicles
UEFA *see* European Football Championships
United Nations 73, 171, 175, 180, 181

unmanned aerial vehicles (UAE) 1, 124, 125, 133
  see also airspace surveillance
urban development 16, 50, 51, 77, 88, 89, 91, 95, 103–16, 137–40, 148, 150, 153–4, 158–9, 162
  festivalisation and 106
  legitimation of 107, 148
  post-Fordist development 105
  see also gentrification, legacies: spatial or infrastructural, militarization: of urban space

Vancouver/Whistler 2010 see Olympic Games
vehicle tracking 1, 11, 44, 62, 93, 98, 139, 143–4, 148, 163
  see also mobility, transit
violence 2, 9, 22, 24, 37, 46, 50, 51, 77, 79, 81, 82, 92, 111, 126, 139, 142
  see also hooliganism, militarization, protest, protest: policing dissent, terrorism

voice analyzers 40, 62
  see also biometrics, face recognition, fingerprinting

WADA see World Anti-Doping Agency
warfare 5, 63, 65, 77–8, 141
  Ares, Olympian god of 28
  see also missile defence, nuclear proliferation, militarization, military-industrial complex
war on terror 22, 73, 125
  see also terrorism, terrorist attacks
watch lists 28, 62, 147
  see also databases, dataveillance, mobility
wiretapping 11, 56, 60, 64–5, 66
  see also communication technology
World Anti-Doping Agency (WADA) 10
  see also drug testing

X-ray technology 40, 128, 129, 140, 142, 143, 144–5

9/11 see terrorist attacks: 9/11